Machine Learning and the Internet of Things in Solar Power Generation

The book investigates various MPPT algorithms, and the optimization of solar energy using machine learning and deep learning. It will serve as an ideal reference text for senior undergraduate, graduate students, and academic researchers in diverse engineering domains including electrical, electronics and communication, computer, and environmental.

This book:

- Discusses data acquisition by the internet of things for real-time monitoring of solar cells.
- Covers artificial neural network techniques, solar collector optimization, and artificial neural network applications in solar heaters, and solar stills.
- Details solar analytics, smart centralized control centers, integration of microgrids, and data mining on solar data.
- Highlights the concept of asset performance improvement, effective forecasting for energy production, and Low-power wide-area network applications.
- Elaborates solar cell design principles, the equivalent circuits of single and two diode models, measuring idealist factors, and importance of series and shunt resistances.

The text elaborates solar cell design principles, the equivalent circuit of single diode model, the equivalent circuit of two diode model, measuring idealist factor, and importance of series and shunt resistances. It further discusses perturb and observe technique, modified P&O method, incremental conductance method, sliding control method, genetic algorithms, and neuro-fuzzy methodologies. It will serve as an ideal reference text for senior undergraduate, graduate students, and academic researchers in diverse engineering domains including electrical, electronics and communication, computer, and environmental.

Smart Engineering Systems: Design and Applications
Series Editor: Suman Lata Tripathi

Internet of Things: Robotic and Drone Technology
Edited by Nitin Goyal, Sharad Sharma, Arun Kumar Rana, and Suman Lata Tripathi

Smart Electrical Grid System: Design Principle, Modernization, and Techniques
Edited by Krishan Arora, Suman Lata Tripathi, and Sanjeevikumar Padmanaban

Artificial Intelligence, Internet of Things (IoT) and Smart Materials for Energy Applications
Edited by Mohan Lal Kolhe, Kailash J. Karande, and Sampat G. Deshmukh

Artificial Intelligence for Internet of Things: Design Principle, Modernization, and Techniques
Edited by N. Thillaiarasu, Suman Lata Tripathi, and V. Dhinakaran

Machine Learning and the Internet of Things in Solar Power Generation
Edited by Prabha Umapathy, Jude Hemanth, Shelej Khera, Abinaya Inbamani, and Suman Lata Tripathi

For more information about this series, please visit: https://www.routledge.com/9781032299785

Machine Learning and the Internet of Things in Solar Power Generation

Edited by
Prabha Umapathy
Abinaya Inbamani
Suman Lata Tripathi
Jude Hemanth and
Shelej Khera

CRC Press
Taylor & Francis Group
Boca Raton London New York

CRC Press is an imprint of the
Taylor & Francis Group, an **informa** business

First edition published 2023
by CRC Press
6000 Broken Sound Parkway NW, Suite 300, Boca Raton, FL 33487-2742

and by CRC Press
4 Park Square, Milton Park, Abingdon, Oxon, OX14 4RN

CRC Press is an imprint of Taylor & Francis Group, LLC

ISBN: 978-1-032-29978-5 (hbk)
ISBN: 978-1-032-29981-5 (pbk)
ISBN: 978-1-003-30296-4 (ebk)

DOI: 10.1201/9781003302964

Typeset in Sabon
by SPi Technologies India Pvt Ltd (Straive)

Contents

Preface to the first edition

Solar is the most reliable source of renewable energy. It is free and available all the time. A vast amount of energy is available, and we would do well to tap into this. The conversion of solar radiation into electricity requires panels to have optimum tilt angle, appropriate ambient temperature, and efficient solar cells; optimizing these factors can maximize energy production. Solar panels should be selected accordingly. This book focuses on the design and development of solar cells by educating readers on solar cell operation and design principles. The Conventional & Non-Conventional SEPIC Converter in Solar Photovoltaic System using Proteus is explained. There is also focus on the design and implementation of the non-inverting buck converter based on the performance analysis scheme. There is also investigation into various solar MPPT techniques. We explore the solar energy forecasting architecture using deep learning model. This book is also intended to serve a fundamental understanding of the characterization of $CuO-SnO_2$ composite nano powder using the hydrothermal method for solar cells. The design and development of solar PV-based advanced power converter topologies for EV fast charging are also covered. The content within chapters is organized and described so as to felicitate readability and application.

Chapter organization

This book is organized into nine chapters:

Chapter 1 presents solar analytics using AWS serverless services.

Chapter 2 studies the Conventional & Non-Conventional SEPIC Converter in Solar Photovoltaic System using Proteus.

Chapter 3 focuses on the design and implementation of non-inverting buck converter based on the performance analysis scheme.

Chapter 4 investigates solar MPPT techniques for solar panels.

Chapter 5 covers real-time solar farm performance monitoring using IoT.

Chapter 6 explores solar energy forecasting architecture using deep learning models.

Chapter 7 provides a fundamental understanding of the characterization of CuO–SnO_2 composite nano powder using hydrothermal methods for solar cells.

Chapter 8 addresses the design and development of solar PV-based advanced power converter topologies for EV fast charging.

Chapter 9 presents the assessment of different MPPT techniques for PV systems.

Editors bios

Prabha Umapathy is currently the Principal at Dr. N.G.P. Institute of Technology Coimbatore. She received her PhD in 2010 from Faculty of Engineering and Technology, Multimedia University, Malaysia. She received her M.E degree in Electrical Machines from PSG College of Technology, Bharathiyar University, Coimbatore in the year 1997, and her BE degree in Electrical & Electronics Engineering from Coimbatore Institute of Technology, Bharathiyar University, Coimbatore in 1993. She has 26 years of teaching experience and around two decades of research experience. She also has ten years of experience in teaching and research done internationally. Her research interests include Power System Optimization, Power System Stability, Power Quality issues, Renewable Energy Resources, Smart Grid and Electric Vehicles. She has around 125 publications to her credit in both international journals and international conferences. She has also visited countries like Malaysia, Singapore, Thailand and Indonesia for her academic and research purposes. She is a recognized PhD supervisor of Anna University, Chennai and seven scholars have graduated under her guidance. Presently she is supervising eight PhD scholars. She has fetched funding from different funding agencies like UGC, AICTE, IEEE, ISTE, etc. for a tune of around 40 Lakhs. She has four copyrights and four patents to her credit. She is the recipient of the International Award, "Distinguished Women in Engineering", instituted by Venus foundation in the field of Electrical and Electronics Engineering.(2017). She has been awarded the "Best Teacher Award", instituted by the Institute of Exploring Advances in Engineering (IEAE-2018).She is a member of IEEE, IEAE, IAEMP and Life member of ISTE.

Abinaya Inbamani is a full-time Research Scholar at Dr. N.G.P. Institute of Technology Coimbatore. She did her Post Graduate degree in Embedded Systems from SASTRA University, Thanjavur. She completed her undergraduate degree in Electrical and Electronics Engineering from Panimalar Engineering College, Chennai. She has a teaching experience of six years. She had been given the Best Teacher Award by the Institute of Scholars. Her area of interests include digital logic circuits, embedded systems and

IoT. She has published papers in various Scopus journals and international conferences. She is also an online educator and also the Coursera beta tester, and is a writer of technologist and SEO content.

Suman Lata Tripathi completed her PhD in the area of microelectronics and VLSI from MNNIT, Allahabad. She did her MTech in Electronics Engineering from UP Technical University, Lucknow and BTech in Electrical Engineering from Purvanchal University, Jaunpur. She is associated with Lovely Professional University as a professor with more than seventeen years of experience in academics. She has published more than 55 research papers in refereed IEEE, Springer and IOP science journals and conferences. She has also published 11 Indian patents and two copyrights, and organized several workshops, summer internships, and expert lectures for students. Other work includes session chair, conference steering committee member, editorial board member, and peer reviewer in international/national IEEE, Springer, Wiley etc Journal and conferences. She received the "Research Excellence Award" in 2019 at Lovely professional university, and received the Best Paper award at IEEE ICICS-2018. She has edited and authored more than 14 books and as well as one book series in different areas of Electronics and electrical engineering. She is associated for editing work with top publishers including Elsevier, CRC Taylor and Francis, Wiley-IEEE, SP Wiley, Nova Science and Apple academic press. She is also associated as an editor of a book series on "Green Energy: Fundamentals, Concepts, and Applications" and "Design and development of Energy efficient systems", to be published by Scrivener Publishing, Wiley (in production). She is also associated with Wiley-IEEE for her multi-authored (ongoing) book in the area of VLSI design with HDLs, and is working as series editor for title, "Smart Engineering Systems" CRC Press Tylor and Francis. She has already completed one book with Elsevier on "Electronic Device and Circuits Design Challenges to Implement Biomedical Applications". Suman is a guest editor for a special issue of "Current Medical Imaging" for Bentham Science. She is associated as senior member IEEE, Fellow IETE and Life member ISC and continuously involved in different professional activities along with academic work. Her areas of expertise include microelectronics device modeling and characterization, low power VLSI circuit design, VLSI design of testing, and advance FET design for IoT, Embedded System Design and biomedical applications.

Jude Hemanth received his BE degree in ECE from Bharathiar University in 2002, ME degree in communication systems from Anna University in 2006, and PhD from Karunya University in 2013. His research areas include computational intelligence and image processing. He has authored more than 150 research papers in reputed SCIE indexed International Journals and Scopus indexed International Conferences. His Cumulative Impact Factor is over 205. He has published 37 edited books with reputed publishers such as Elsevier, Springer and IET, and serves as Associate Editor/Scientific Editor of SCIE Indexed International Journals such as IEEE Journal of Biomedical and Health Informatics (IEEE-JBHI), Soft Computing (Springer), Journal of Intelligent and fuzzy systems, Mathematical Problems in Engineering. IET Quantum Communications and Dyna (Spain). He serves as an editorial board member/guest editor of many journals with leading publishers including Springer, Inderscience, MDPI, and IGI Global. He is the series editor of the book series "Biomedical Engineering" in Elsevier and "Robotics & Healthcare" with CRC Press.

He has received a project grant of £35,000 from the UK Government (GCRF scheme) with collaborators from University of Westminster, UK. He has also completed a funded research project for CSIR, Government of India, and worked on an ongoing funded project from DST, Government of India. He also serves as the "Research Scientist" of Computational Intelligence and Information Systems (CI2S) Lab, Argentina; LAPISCO research lab, Brazil; RIADI Lab, Tunisia and Research Centre for Applied Intelligence, University of Craiova, Romania

He has been also the organizing committee member of several international conferences across the globe such as Portugal, Romania, UK, Egypt, China, etc. He has delivered more than 100 keynote talks and invited lectures in International conferences and workshops. He holds professional membership with IEEE Technical Committee on Neural Networks (IEEE Computational Intelligence Society) and IEEE Technical Committee on Soft Computing (IEEE Systems, Man and Cybernatics Society) and ACM. Currently, he is working as Associate Professor in the Department of ECE, Karunya University, Coimbatore, India. He also holds the position of Visiting Professor in Faculty of Electrical Engineering and Information Technology, University of Oradea, Romania.

Shelej Khera completed his PhD in the area of Wireless Communication in Electronics & Communication engineering at Lingaya's University, Faridabad, under the guidance of senior professor Dr. SVAV Prasad. The external examiner for open seminar was Dr. Manav Bhatnagar Professor from IIT Delhi & external examiner for final defense was Dr. D. R. Bhaskar Sr. Professor ECE DEPTT. Jamia Millia Islamia University, New Delhi. He did his MTech in Electronics & Communication Engineering from Punjab Technical University, Jalandhar and BTech in Electronics & Telecommunication Engineering from Pune University, Poona. He is associated with Lovely Professional University as a Professor with more than twenty-five years of experience in academics. He has worked as Dean Academics UG & PG, Professor & HOD (ECE), MTECH (ECE) coordinator, NBA coordinator, Management Representative (MR) ISO, in charge Entrepreneurship & development cell in various colleges. He has published more than 40 research papers in refereed IEEE, Springer and IOP science journals and conferences. He has guided more than 25 MTech theses and is presently guiding seven PhD research scholars. He has received various research grants from AICTE, ISTE, DST and EDI. He has organized several workshops, summer internships, and expert lectures for students. He received the Best Paper award at IEEE ICICS-2018. He has edited five books and a book series in different areas of electronics and electrical engineering. His area of expertise includes wireless communication, next generation wireless technologies, channel estimation, 5G technologies, artificial intelligence, neural networks.

Contributors

N. Divya
Sri Ramakrishna Engineering College
Coimbatore, Tamil Nadu, India

R. Divya
PSG Institute of Technology and
  Applied Research
Coimbatore, Tamil Nadu, India

Agam Das Goswami
VIT-AP University
Amravati, Andhra Pradesh, India

Parul Dubey
Dr. C.V. Raman University
Chhattisgarh, India

N.R. Govinthasamy
Dr. N.G.P. Institute of Technology
Coimbatore, Tamil Nadu, India

S. Jaganathan
Dr. N.G.P. Institute of Technology
Coimbatore, Tamil Nadu, India

E. Kalaivani
Bannari Amman Institute of
  Technology
Erode, Tamil Nadu, India

E. Kannapiran
Dr. N.G.P. Institute of Technology
Coimbatore, Tamil Nadu, India

M. Karthik
Sri Ramakrishna Engineering College
Coimbatore, Tamil Nadu, India

S.S. Karthikeyan
Dr. N.G.P. Institute of Technology
Coimbatore, Tamil Nadu, India

C. Kathirvel
Sri Ramakrishna Engineering
  College
Coimbatore, Tamil Nadu, India

S. Kavitha
Gobi Arts & Science College
Gobichettipalayam, Tamil Nadu,
  India

R. Mohan Kumar
Sri Ramakrishna Engineering
  College
Coimbatore, Tamil Nadu, India

C.V. Pavithra
Assistant Professor (Sr. Gr)
PSG Institute of Technology and
  Applied Research
Coimbatore, Tamil Nadu, India

V. Radhika
Sri Ramakrishna Engineering
  College
Coimbatore, Tamil Nadu, India

R. Felshiya Rajakumari
Karpagam Academy of Higher
    Education
Coimbatore, Tamil Nadu, India

Shreyas Rajendra Hole
VIT-AP University
Amravati, Andhra Pradesh, India

M. Siva Ramkumar
Karpagam Academy of Higher
    Education
Coimbatore, Tamil Nadu, India

R.R. Rubia Gandhi
Sri Ramakrishna Engineering College
Coimbatore, Tamilnadu, India

V. Rukkumani
Sri Ramakrishna Engineering College
Coimbatore, Tamil Nadu, India

M. Saravanakumar
Gobi Arts & Science College
Gobichettipalayam, Tamil Nadu,
    India

K. Srinivasan
Sri Ramakrishna Engineering
    College
Coimbatore, Tamil Nadu, India

M. Sundaram
PSG College of Technology
Coimbatore, Tamil Nadu, India

R. Uthirasamy
Mahendra Engineering College
Namakkal, Tamil Nadu, India

Arvind Kumar Tiwari
Dr. C.V. Raman University
Chhattisgarh, India

Chapter 1

Solar analytics using AWS serverless services

Shreyas Rajendra Hole and Agam Das Goswami
VIT-AP University, Amravati, Andhra Pradesh, India

CONTENTS

1.1 INTRODUCTION

The magnificent and limitless amount of pure, renewable energy provided by the sun may be enjoyed by all of the world's people. As a matter of fact, in a single hour, our planet absorbs more solar energy than the whole globe uses in a year. Solar photovoltaic (PV) modules (photo = light, voltaic = electricity) have the potential to convert the energy emitted by the sun into electrical current.

In general, solar technology may be divided into the following types: Photovoltaic (PV) systems, concentrated solar electricity, and solar water heating are all examples of active solar approaches. The drying of garments and warming of the air are two examples of active solar power in action. A few examples of passive solar approaches include utilizing solar orientation, high thermal mass or light-dispersing materials, and design features that promote natural airflow into a structure.

Over the course of a month, the world's supply of fossil fuels is about equal to the quantity of solar energy that arrives on Earth. As a result, solar

DOI: 10.1201/9781003302964-1

energy has a worldwide potential that is many folds more than the total energy required of the whole globe at the moment. There are a number of technical and economic challenges to overcome before widespread use of solar energy is even considered. It will be our capacity to overcome scientific and technical hurdles, together with marketing and financial issues, along with political and regulatory considerations, such as renewable energy tariffs, that will define the future of solar power installations.

1.1.1 Solar thermal power

Photovoltaics (PV), concentrated solar power (CSP), and both are all types of solar power that use the sun's free energy to make electricity. Concentrated solar power systems use lenses or mirrors and solar tracking technology to focus sunlight into a small beam that can be used for electricity. The photovoltaic effect is used in solar cells to turn light energy into electricity [1].

A single solar cell-powered calculator or an off-grid rooftop PV installation may power a remote dwelling. Photovoltaics, on the other hand, have largely been utilized as a power source for small and applications with a medium-sized scope up until recently. In the 1980s, CSP plants became financially feasible. In recent years, as the price of solar electricity has reduced, the number of grid-connected solar PV systems has risen exponentially. Gigawatt-scale solar power plants and millions of smaller ones have been constructed. It has become a low-cost, low-carbon technology in recent years. IEA's "Net Zero by 2050" scenario forecasts that solar power will contribute around 20 percent of global energy consumption by 2050, making it the biggest source of electricity in the world.

In terms of solar installations, China has the most. In 2020, solar power will account for 3.5% of global energy production, up from less than 3% in the previous year. Utilities could anticipate to spend around $36/MWh for utility-scale solar power in 2020 [2], with installation costing about one dollar per DC watt [3].

Several rich nations have incorporated huge volumes of solar energy capacity into their power systems in order to add to or substitute for existing energy resources. Meanwhile, developing nations are increasingly turning to solar energy to reduce their reliance on expensive imported fuels and to reduce their carbon footprint. The utilization of long-distance transmission allows for the replacement of fossil fuels with renewable energy resources located a great distance away. Solar power plants employ one of two techniques to generate electricity. First is Photovoltaic (PV) systems, whether mounted on a roof or installed on the ground, capture and convert sunlight directly into power. Second is Concentrated solar power (CSP) facilities create electricity by using solar heat to generate steam, which is then pushed via a turbine to generate electricity.

1.2 RELATED WORK

In order to increase the usage and expansion of thermal energy in an ecologically sustainable manner, weather conditions are forecast, enhancing the efficiency and consistency of energy distribution over the long term [5]. Maintenance of solar power plants is carried out via the execution of operations and maintenance (O&M) activities that make use of predictive analytics, supervisory control, and data collection in order to boost and improve performance (SCADA). The utilization of the internet (cloud) and Internet of Things (IoT) devices, as well as operations and maintenance, supervisory control, and data acquisition, may improve preventive maintenance. There is a considerable increase in the efficiency of solar power plants in India as a result of efforts by a variety of various industries across the nation. The goal of this article is to increase the efficiency of solar power plants via the use of the IoT and predictive analytics.

The authors of a research paper employed three free datasets given by National Institutions to investigate historical trends and policy alternatives for soil usage and power consumption, as well as to evaluate correlations between these variables [7]. It was decided to undertake the analysis on a provincial level. Thereafter it was concluded that the deviations from the policy scenarios needed to be addressed in order to emphasize the need for policy ideas and pathways to achieve the goal of a renewable energy share and a decrease in SO_2 consumption trends by 2030. For example, because of its position in the field of renewable energy integration, building integrated photovoltaics (BIPV) is considered to be a critical driver in the solution of this issue.

Photovoltaic (PV) systems and electrical energy storage (EES) for PV systems are both explored in length in a research paper [8]. Cellular technologies of the future are examined. The study also involves a look at solar power forecasting methodologies for PV and EES operation and planning. Sizing PV and energy storage systems with anaerobic digestion biogas power plant (AD) is established in order to avoid energy imbalance between production and demand due to AD generator limits as well as high penetration of photovoltaic (PV) panels. Correlation analysis in machine learning faces challenges from uneven data and data uncertainty. With the support of extensive assessments of real solar irradiance and meteorological condition data, a prospective framework has been built and tested. To construct the clearness index (CI), a cluster analysis utilizing Fuzzy C-Means with dynamic temporal warping and other methodologies is done on real-life solar data.

The quantity of solar radiation that enters the building has a considerable impact on the design of solar architecture. It is the building envelope that is most important when it comes to solar heat and light transmission since it determines how well the structure performs [9]. In the development of solar architecture, the use of computer modeling and simulation is essential. When doing this kind of computer analysis, it is common to utilize a year's worth of weather data that has been saved in a typical weather file. The total of ultraviolet, visible, and near-infrared (UV, VIS, and NIR) light is used in typical solar architecture design analytics; nevertheless, these three components play distinct roles in a building's energy efficiency. It may be beneficial to analyze the different solar modules in isolation from one another. With the use of freely available information from ground weather stations, models for the estimation of visible and near-infrared components will be created that can be easily integrated into current available solar architecture design and topics of research activities. The classification-based modeling techniques were used to examine and assess the decomposition of hourly worldwide horizontal solar VIS and NIR components from traditional weather files into hourly global horizontal solar VIS and NIR components from traditional weather files. An established technique for using these models for solar architecture design and analysis has been developed, which is described in detail in the research article [9].

Another investigation with a purpose to find out Uttarakhand's solar potential in the cloud using Perl and Geographic Information Systems (GIS) [10]. The Citrix XenCenter hypervisor, which is a bare-metal hypervisor, was used to develop the cloud computing environment. According to Pyranometer and the National Renewable Energy Laboratory website, some GHI data has been collected. The GHI of the sloped surface was transformed using Perl programming in order to guarantee that the solar PV got the maximum amount of solar irradiation feasible for its size. These solar resource maps may be useful for a variety of purposes, including PV installations, resource planning, solar savings, energy trading, and subsidies. From people to organizations and the federal government, they are all accessible.

By constructing solar power plants, one may alter the landscape in a positive way. This landscape alteration has sparked concerns about the aesthetic impact, land-use competition, and the end-of-life stage of solar power facilities, among other things. Existing research [11] advises connecting solar power plants with their surrounding landscapes, a concept known as the solar landscape, in order to address these challenges. The goal of solar landscapes is to give a variety of benefits in addition to electricity generation, such as decreasing visibility and establishing habitats, however there is a paucity of scientific study on solar landscapes today. It is our aim that this comparative analysis of 11 prominent occurrences will assist us in better understanding solar landscapes by assessing their visibility (as well as their multifunctionality and temporality). Every situation results in lowered sight. However, there are five occasions in which recreational facilities may assist in increasing visibility. The cases performed between six and fourteen roles in terms of providing, regulating, and cultural expression. Functions might be found in arrays, between arrays, and in the vicinity of solar patches. The majority of the time, concerns about the future use of the sites were raised, but only in two cases were new landscape components introduced to encourage the future use of the areas. The other issues were addressed in other ways. In this case study [11], many concerns, such as the aesthetic impact, land-use competition, and the end-of-life stage, are explored in further detail. Taking them all together, they provide a diverse variety of approaches to addressing societal concerns, but the full potential of three crucial characteristics has yet to be realized. This comparative research also demonstrates the need of addressing the development of trade-offs between spatial elements as well as the necessity to discriminate between different types of solar landscapes. Using the analytical technique that has been developed, it is feasible to analyze both the good and negative features of solar power installations.

It has been proposed in a study to simulate the solar power generating system in Medellín, Colombia, with the use of ML technologies [12]. There are four forecasting models that have been created utilizing approaches that are compatible with ML and AI technologies. They are the K-Nearest Neighbor (KNN), Linear Regression (LR), Artificial Neural Networks (ANN), and Support Vector Machines (SVM) (SVM). The accuracy of the four techniques utilized to estimate solar energy output was determined to be high. The ANN forecasting model, on the other hand, delivers the best accurate prediction, according to the RMSE and MAE. Medellín's PV power production was predicted using ML models, and it was discovered that these models were both possible and successful.

1.3 INTERNET OF THINGS (IOT)

The IoT allows objects that have electronics incorporated into their design to communicate and experience interactions with one another and with the outside world via the use of wireless technology (IoT). During the next several years, the IoT has the potential to completely transform the way individuals go about their daily lives. The IoT has achieved important advances in a variety of disciplines, including medicine, energy, gene therapy, agriculture, smart cities, and smart homes, among others [13]. It is estimated that more than 9 billion "Things" (physical goods) are now available for purchase on the internet. In the not-too-distant future, this number is expected to soar to an astounding 20 billion individuals. Figure 1.1. shows an IoT based solar monitoring system.

For IoT, there are four primary components:

1. Embedded systems with low power consumption.
2. Computing in the cloud.
3. Readily available big data.
4. A link to the internet.

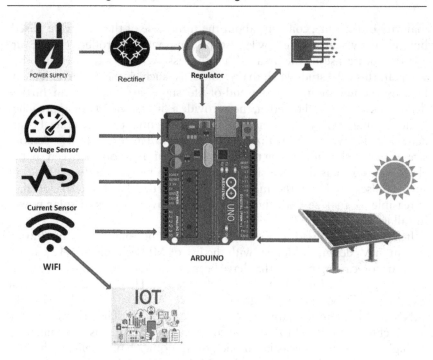

Figure 1.1 Solar energy monitoring system powered by IoT.

As a consequence of the Internet of Things' rising bandwidth and lowering hardware prices, the surroundings of many industries are rapidly changing. There are several new applications for the IoT being developed in a variety of industries such as healthcare, construction, as well as government and insurance. As a consequence of these advancements, large corporations, financial institutions, and other organizations are all increasing their investment on information technology.

The Industrial IoT includes a wide range of device, application, and engineering systems, but they all have the same fundamental components. IoT architecture is explained in Figure 1.2.

- **Smart gadgets and sensors**
  The top layer of the protocol stack consists of connections between devices and sensors. The smart sensors are continually collecting and transmitting data from their immediate surroundings, which they refer to as the immediate environment. It is now feasible to manufacture tiny smart sensors for a broad variety of different applications by using the most up-to-date semiconductor technology available on the market.

  First and foremost, information about their immediate surroundings is obtained by sensors and devices that might vary from a simple temperature measurement to a full live video stream. The phrase "sensor/

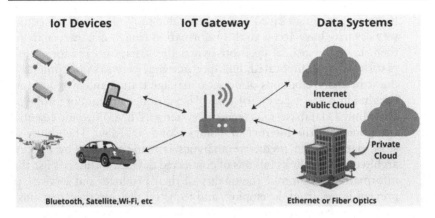

Figure 1.2 Architecture of IoT.

device" may be used to describe a sensor that is a component of a larger gadget that performs functions other than just sensing the surrounding environment. When it comes to sensors (such as a camera and an accelerometer), phones are more than just sensors, since they are capable of performing a broad range of functions with their sensors.

It is necessary to set up a cloud service in order for data to be transferred. Connecting sensors and devices to the cloud may be accomplished via the use of cellular, satellite, Wi-Fi, Bluetooth, low-power wide area networks (LPWAN), a gateway, or a router. Considering power usage, range, and bandwidth are all important considerations when making a selection. Connection choices for IoT applications are many, but all are aimed on transmitting data to a cloud service.

- Gateway
  The IoT Gateway is responsible for coordinating bidirectional data flow across a variety of networks and protocols. Gateways are devices that translate between different network protocols in order to guarantee that linked devices and sensors are compatible with one another. Pre-processing of data gathered from hundreds of sensors may be conducted locally by gateways before it is forwarded to the next level. Because of the compatibility of the TCP/IP protocol, it may be necessary in particular situations. Higher-order encryption algorithms, when used in conjunction with IoT gateways, provide a certain level of security for the network as well as the data that is sent over the network, according to the FBI. It protects the system against damaging assaults and unauthorized access by acting as an intermediary layer between the cloud and the devices.

- Cloud
  Devices, applications, and users connected to the IoT will create tremendous volumes of data, and managing this data will be a significant issue. On real time, data can be collected and analyzed, while also

being managed and saved in the cloud, due to the IoT. Businesses and services may have access to this information from a distance, enabling them to make critical decisions when the situation calls for it. IoT platforms are complicated, high-performance networks of computers that can analyze billions of devices, manage traffic, and give accurate insights in a short period of time. They are becoming more popular. Distributed database management systems (DBMS) are an essential component of the Internet of Things cloud (DBMSs). The ability to store data and do predictive analytics is made possible by a cloud architecture that links billions of connected devices. Businesses use this information to improve the quality of their products and services, to prevent issues from developing, and to better develop their new business model based on the information they have accumulated.

1.4 THE USE OF CONTEMPORARY TECHNOLOGIES IN SYSTEM DESIGN

Time and technology improvements have resulted in a rush of new tools during the previous decade. These instruments have a positive impact on systems and processes. Some of the areas and technologies that may be employed to increase performance are mentioned below. Figure 1.3 displays the most frequent new system design tools. It highlights the interconnectedness between AI, ML, and DL. When it comes to artificial intelligence (AI), machine learning (ML) is the subset and when we discuss ML, Deep Learning (DL) is a subset to ML.

- Artificial Intelligence
  It is a wide topic in the study of artificial intelligence (AI), which is divided into many subfields, that computers' ability to develop logical

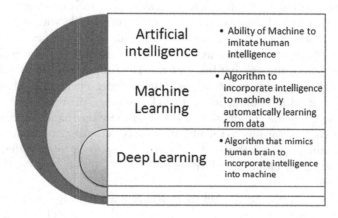

Figure 1.3 Relationship between AI, ML, DL.

behavior in response to external inputs is a broad issue in AI. The ultimate goal of artificial intelligence will be the development of systems that can do tasks that would otherwise need human intelligence. There are many different perspectives on AI and how it relates to the products and services that we use on a daily basis. One of the project's objectives is to develop expert systems that can learn from their users, exhibit their capabilities, explain their capabilities, and give advice. The ultimate goal of infusing human intelligence into robots is to create machines that are capable of comprehending, reasoning, and learning in the same way that humans are capable of doing these things. Figures 1.4 and 1.5 demonstrates a broad spectrum of artificial intelligence applications (AI). Figure 1.4 describes types of AI on the basis of technology, whereas Figure 1.5 focuses on types of AI on the basis of functionality.

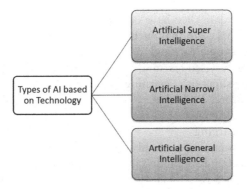

Figure 1.4 AI classification on the basis of technology.

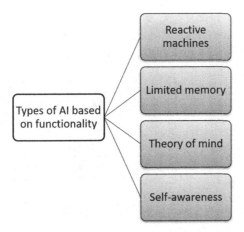

Figure 1.5 AI classification on the basis of functionality.

In recent years, as machine learning has advanced during the preceding two decades, there has been an increase in interest in AI. It has become more popular in the twenty-first century to use artificial intelligence in many applications (AI). Consequently, ML may be utilized to construct systems that are always improving.

- **Machine Learning**
  Computers perform algorithm-based procedures, which means there is no space for mistake in their results. Data from current samples may be utilized in place of written instructions that produce a response based on the data provided, enabling computers to make decisions based on the data. Computers, like people, have the capacity to make mistakes when making judgments, and this is no exception. As a result, it is a kind of ML in which computers can learn in the same way that people do, by using data and previous experiences [26]. One of the fundamental goals of ML is to develop prototypes that can self-improve, recognize patterns, and propose answers to new problems based on the data they have collected in the past. Figure 1.6 explains the types of ML. ML may be broken down into five categories:
  1. Supervised learning.
  2. Unsupervised learning.
  3. Semi-supervised learning.
  4. Reinforced learning.
  5. Deep learning.

- **Supervised learning**
  A dataset containing some observations and their labels/classes must exist for this family of models in order for them to be used. For example In order to perform observations, it may be necessary to utilize species photos, with the labels identifying each animal by its scientific name.

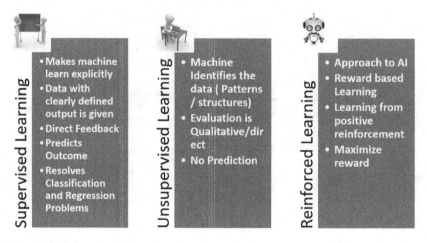

Figure 1.6 Types of machine learning.

First and foremost, these models must be trained by learning from the labeled datasets before they can make predictions about the future. Whenever new, unlabeled data are added to the model, the learning algorithm makes use of the inferred function that it has created in order to anticipate what will happen when the new observations are received. In the wake of proper training, a model is capable of generating goals for each new data it encounters. Additionally, the learning algorithm may detect errors by comparing its output to the intended output (the ground truth label) and changing its output as required to get the desired result (e.g., via back-propagation). Regression modeling and classification modeling are the two types of supervised modeling approaches. In the former, we make predictions or draw conclusions based on the facts already available. In the latter, an analysis method known as categorization divides large datasets into smaller subsets based on common characteristics is used.

- **Unsupervised learning**
  It is necessary to have some data in order to make use of this family of models; however, it is not necessary to know the labels or classifications of the data in advance. If a system intends to infer a function from unlabeled data, the data must first be labeled in order to explain a hidden structure that has been discovered. In unlabeled data, hidden structures may be defined by employing a data exploration technique and inferences derived from datasets, and a system can make use of data exploration and inferences made from datasets in order to do this.

- **Semi-supervised learning**
  This topic is conveniently located in the midst of both the supervised and unsupervised learning groups, making it simple to cover all of the material. Semi-supervised models learn on data that is both labeled and unlabeled, and this data is used throughout the training phase. In situations when the quantity of labeled data is less than the amount of unlabeled data, both unsupervised and supervised learning are rendered worthless. Many instances exist in which material that has not been properly categorized is exploited to make generalizations about persons without their knowledge. This strategy is referred described as "semi-supervised learning" in generic terms, although it is more specific. Semi-supervised learning, as opposed to unsupervised learning, makes use of a labeled data set rather than a randomly produced data set. Similarly to supervised learning, the labeled data contains more information than the data that must be predicted. This type of learning makes use of smaller labeled datasets than projected datasets, which makes it more efficient.

- **Reinforced learning**
  These algorithms utilize estimated mistakes as incentives or penalties in order to reward or penalize its users. Those found guilty of a

significant mistake face severe penalties and get little compensation. There will be no harsh penalties or significant rewards if the fault is minor.

- **Deep learning**

  It is the only kind of machine learning that focuses on training the computer to replicate human behavior, which is known as deep learning. Deep learning is the process through which a computer system learns to identify complicated information, such as photographs, text, or speech, via trial and error. For example, this method may be able to attain accuracy levels comparable to those of the current state of the art (SOTA). In certain circumstances, it may even exceed humans in terms of performance. They must be trained using a large quantity of labeled data and neural network topologies with several layers in combination, both of which are time-consuming. DL has the following noteworthy characteristics:

  - Developing technology such as virtual assistants, facial recognition, driverless cars, and other comparable technologies relies heavily on Deep Learning.
  - It begins with the collection of training data and ends with the use of the training's outputs.
  - To describe the learning technique, the term "deep learning" has been adopted since the neural networks quickly learn about new layers of data that have been introduced to the dataset over the course of a few minutes. To improve the system as a whole, all training is done with the goal of making it more efficient.
  - It has been widely accepted as a training and deep learning system by data experts, which has resulted in a significant increase in training efficiency and power.

1.5 SERVERLESS SOLAR DATA ANALYTICS

There are a variety of smart devices and sensors that can be used to create raw data, which can then be analyzed in more detail using analytics software. Smart analytics solutions are required for the administration and optimization of IoT systems. When an IoT system is correctly built and implemented, it is feasible for engineers to discover irregularities in the data gathered by the system in real time and to take quick action in order to avoid a negative consequence that may have occurred otherwise. If the data is gathered in the proper manner and at the appropriate time, service providers will be able to prepare for the next step. Large corporations are taking use of the massive volumes of data created by IoT devices in order to get fresh insights and open up new business opportunities. It is feasible to foresee market trends and prepare for a successful implementation via the careful use of market research and analytical techniques. Predictive analysis, which is a key component of every business model, has the potential to significantly improve a company's capacity to perform in its most vital areas of business.

We will make use of Amazon Web service (AWS), a well-known cloud service provider, in order to complete data analytics serverless. The architecture used to obtain the analytics is depicted in Figure 1.7. Assume we have an N-number of Internet-of-Things devices/sensors that are broadcasting the intensity of sunlight to the cloud in real time. The rest of the analytics will be handled by this architecture, which will be serverless in nature.

- **Kinesis Data Streams**
  Amazon Kinesis Data Streams is a streaming data service that is completely controlled by Amazon. In addition to click streams and application logs, Kinesis streams may receive data from hundreds of thousands of sources at any same time [27]. Kinesis Applications will be able to access the data in the stream and begin processing it within a few seconds of the stream being created. Amazon Kinesis Data Streams takes care of all of the Hardware, memory, connection, and configuration concerns necessary to stream your data at the pace you choose. Everything from provisioning to deployment to ongoing maintenance and other services for your data streams is handled on your behalf. The synchronous replication of data across three AWS Regions provided by Amazon Kinesis Data Streams ensures high availability and long-term data durability.

  We may begin utilizing Kinesis Streams after signing up for Amazon Web Services by following these steps:
  1. Create a Kinesis stream; this can be done through either Management Console or CreateStream operation.
  2. Constantly generating new data from our data generators.
  3. Using the Kinesis Data Streams API or the Kinesis Client Library stream (KCL), we may create Kinesis applications that read and process data.

- **Kinesis Data Firehose**
  Amazon Kinesis Data Firehose is the most efficient method of transferring streaming data into data repositories and analytics tools, according to the company. As a streaming data storage service, Amazon S3 may be used to collect, transform, and store data streams, which can then be analyzed in near-real time using the business intelligence tools and dashboards that we are currently familiar with. This is a fully managed service that adjusts automatically to the flow of our data and really doesn't need further management on our behalf [28]. It may also batch, compress, and encrypt the data before loading it, which decreases the size of storage space necessary at the destination while increasing security at the same time.

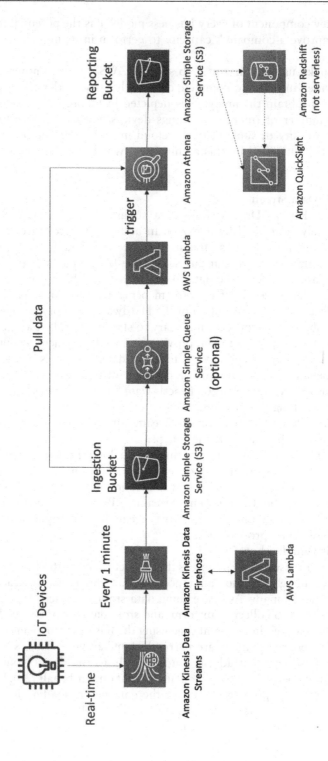

Figure 1.7 AWS severless data analytics solution for solar power.

Customers may gather and load their data into Amazon S3, Amazon Redshift, or Amazon Open Search Service using Amazon Kinesis Data Firehose without having to worry about any of the underlying infrastructure, storage, networking, or setup requirements. We don't have to worry about setting up hardware or software, or about building any new programs in order to keep track of this operation since it is automated. Firehose is likewise scalable, and it does so without the need for any developer engagement or overhead to be involved. This results in the synchronous replication of data across three Amazon Web Services regions, assuring high availability and durability of the data as it travels to its final destinations [31–35].

We may begin utilizing Kinesis firehose after signing up for Amazon Web Services by following the steps:

1. Use the Firehose Console or the Create Delivery Stream operation to create an Amazon Kinesis Data Firehose delivery stream. Optionally, we can set up our delivery stream to use an Amazon Lambda function to preprocess and alter the raw data before importing it.

2. Configure the data generators to send data to your delivery stream on a regular basis by utilizing the Amazon Kinesis Agent or the Firehose API to provide data to the delivery stream.

3. Firehose uploads the data in a continuous and automated manner to the destinations we choose.

- **Amazon S3**
"S3" stands for Simple Storage Service. In order to store and retrieve any amount of data from anywhere, Amazon S3 was designed as an object storage system. An easy-to-use, low-cost solution that enables industry-leading scalability at a fraction of the expense. Using the use of Amazon S3, users may store and access a limitless amount of data at any time and from any location, all through a simple web service interface. This service simplifies the process of developing cloud-native storage-aware applications [29]. One may start small and expand their application as they find appropriate, all while retaining uncompromised speed and reliability because of the scalability of Amazon S3 [36–41].

Furthermore, Amazon S3 is designed to be very versatile. Whether you wish to preserve a little amount of data for backup purposes or a huge amount for disaster recovery, anyone may do so using a simple FTP application or a complicated web platform such as Amazon.com. Amazon S3 relaxes developers from having to worry about where to store their data, allowing them to spend more time innovating.

With Amazon S3, users only pay for the amount of storage space that they consume. There is no predetermined pricing for this service. Your monthly expenditure may be estimated with the aid of the AWS Pricing Calculator. When our overhead is smaller, we may charge a lesser rate. Some items are priced differently in various Amazon S3 regions. The location of your S3 bucket has an impact on the cost of your monthly payment. No data transfer fees are charged for COPY requests that move data inside an Amazon S3 Region. Depending on which AWS region the data is being transmitted to, Amazon S3 charges a fee for each gigabyte of data that is carried across the service. Transferring data from one AWS service to another within the same geographical area, such as the US East (Northern Virginia) region, will not result in a data transfer fee being applied to the account.

The service is completely free, and there is no obligation to pay anything up front or make any commitments before beginning to use it. At the end of each month, users are invoiced for the amount of energy used during the previous month. Tap "Billing and Cost Management" under "Web Services Account" after logging into the Amazon Web Services account to get a list of current spending. We may get started with Amazon S3 for free by taking advantage of the AWS Free Usage Tier, which is available in all regions with the exception of the AWS GovCloud Regions. Furthermore, new AWS customers get 20,000 get requests and 2,000 put requests per month for the first year, as well as 15 GB of data transfer out of Amazon S3 during that time period. The amount of monthly use that is not utilized will not be carried over to the next month.

- **Amazon Athena**
  Amazon Athena is a data processing tool that simplifies the process of examining data stored in Amazon S3 using normal SQL queries. It is available for both Windows and Mac users. It is available for download and usage by users of both Windows and Linux operating systems. It is not necessary to set up or maintain any infrastructure since Athena is a serverless platform, and users may start analyzing data immediately after installing it. We don't even have to input the data into Athena since it works directly with data stored on Amazon S3, so there's no need to do so at all. Begin by login into the Athena Management Console and setting the schema that users will be using. After that, begin querying the data stored in the database by using the Athena API [30]. It is possible to use Athena as a database on the Presto platform, and it includes all of the usual SQL features. Other prominent data formats supported by the tool include Oracle Relational Catalog, Apache Parquet, and Avro. CSV and JSON are only two of the formats supported by the tool. Achieving complicated analysis, despite the fact that Amazon Athena is suitable for fast, ad hoc querying and that it integrates well with Amazon QuickSight for simple visualization, is possible with it. This includes huge joins as well as window functions and arrays, among other things.

  The Amazon Athena query service, the Amazon Redshift data warehouse, and the sophisticated data processing framework, the Amazon EMR, are all intended to suit a broad variety of data processing needs and use case situations, respectively. All that is necessary is that we choose the instrument that is best appropriate for the task at hand. AWS Redshift is the best choice for corporate reporting and business intelligence workloads when we need quick query performance for complicated SQL queries, which is exactly what we need. In addition, it is the most cost-effective option available. With Amazon EMR, it is simpler and less costly to manage widely dispersed processing frameworks like as Hadoop, Spark, and Presto, while also lowering the total cost of ownership. To meet analytic requirements, users may run bespoke apps and code on Amazon EMR. We can also customize particular CPU, memory, storage, and application characteristics to meet our needs. Amazon EMR is a service provided by Amazon. Using Amazon Athena to query data in S3 eliminates the need for any additional servers to be installed or maintained.

- **Amazon QuickSight**
  It is simple for everyone in your organization to get a better understanding of the data thanks to the interactive dashboards, machine learning-powered patterns, and outlier identification in Amazon QuickSight. A cloud-based business intelligence (BI) tool for large organizations. Using Amazon QuickSight, users can easily share insights with their team, no matter where they are located in the world. With Amazon QuickSight, customers can access and aggregate the cloud-based data in a single place. All of this information is accessible in a single QuickSight dashboard, which includes data from AWS, spreadsheets, SaaS, and business-to-business transactions (B2B). Amazon QuickSight is a cloud-based service that is totally managed and provides enterprise-grade security, global availability, and built-in redundancy. It is available on a subscription basis. One may also scale from 10 to 10,000 members using the user management features provided by this solution, all without the need to invest in or maintain any infrastructure.

  In order to better understand and digest data, QuickSight offers decision-makers with an interactive visual environment in which to do so. Dashboards may be viewed securely from any computer or mobile device connect to the network, including a smart phone.

- **Lambda**
  This service allows the user to access their code without the need to set up or maintain servers on own computer. When the code is not being executed, users are only charged for the compute time that was used. Lambda may be used to operate any and all of existing apps and back-end services, with no need for further management. Lambda takes care of everything users need to run and expand the code with high availability, and we just need to submit the code once to take advantage of this. Our code may be activated on demand by an AWS service, a web page, or a mobile application, or we may call it directly.

- **Amazon Simple Queue Service (SQS)**
  A fully managed message queuing service, Amazon Simple Queue Service (SQS) may be used by microservices, distributed systems, and serverless applications, among other things. SQS relieves developers of the responsibilities of maintaining and administering message-oriented middleware, allowing them to devote their time and energy to developing new and innovative functionality. SQS eliminates the need to be concerned about messages being lost or depending on other services to be accessible when you use it.

  In SQS, there are two different types of message queues. Standard queues enable the completion of the highest feasible amount of work due to the best-effort ordering and at-least-once delivery given by standard queues. The use of SQS FIFO technology in the construction of queues ensures that each message is treated only once, in the exact order in which it was given.

1.6 CONCLUSION

The importance of solar energy has been highlighted, and some historical research efforts have been appraised. Real-time solar analytics is essential for improving the performance of solar cells and the structures that are linked to them. Use of Amazon Web Services in combination with IoT devices has the potential to increase efficiency. Although Amazon Web Services (AWS) uses deep learning algorithms to do calculations, the algorithms are self-managed. The serverless architecture has been presented in detail in this chapter.

REFERENCES

[1] Solar | Department of Energy. (n.d.). Energy.gov. https://www.energy.gov/solar
[2] Levelized Cost of Energy and of Storage 2020. (2020, October 19). Lazard.com. https://www.lazard.com/perspective/lcoe2020
[3] Developers Increasingly Pair Batteries with Utility-Scale Solar to Combat Declining Value in Crowded Markets | Utility Dive. (2021, October 13). Utility Dive. https://www.utilitydive.com/news/developers-increasingly-pair-batteries-with-utility-scale-solar-to-combat-d/608117/
[4] Abdou, N., Mghouchi, Y. E., Hamdaoui, S., Asri, N. E., & Mouqallid, M. (2021). Multi-objective optimization of passive energy efficiency measures for net-zero energy building in Morocco. *Building and Environment*, 204, 108141.
[5] Begum, S., & Banu, R. (2018). Improving the performance of solar power plants through IOT and predictive data analytics.
[6] Begum, S., Banu, R., Ahamed, A., & Parameshachari, B. D. (2016, December). A comparative study on improving the performance of solar power plants through IOT and predictive data analytics. In *2016 International Conference on Electrical, Electronics, Communication, Computer and Optimization Techniques (ICEECCOT)* (pp. 89–91). IEEE.

[7] Mancini, F., & Nastasi, B. (2020). Solar energy data analytics: PV deployment and land use. *Energies*, 13(2), 417.

[8] Lai, C. S., Lai, L. L., & Lai, Q. H. (2021). Data analytics for solar energy in promoting smart cities. In *Smart Grids and Big Data Analytics for Smart Cities*, 173–263.

[9] Duan, Q., Feng, Y., & Wang, J. (2021). Clustering of visible and infrared solar irradiance for solar architecture design and analysis. *Renewable Energy*, 165, 668–677.

[10] Kapoor, M., Garg, R. D., & India, V. L. *Solar Energy Planning Using Geospatial Techniques and Big-Data Analytics*.

[11] Oudes, D., & Stremke, S. (2021). Next generation solar power plants? A comparative analysis of frontrunner solar landscapes in Europe. *Renewable and Sustainable Energy Reviews*, 145, 111101.

[12] Gutiérrez, L., Patiño, J., & Duque-Grisales, E. (2021). A comparison of the performance of supervised learning algorithms for solar power prediction. *Energies*, 14(15), 4424.

[13] Introduction To Internet Of Things (IoT) | Set 1 – GeeksforGeeks. (2018, August 14). GeeksforGeeks. https://www.geeksforgeeks.org/introduction-to-internet-of-things-iot-set-1/

[14] Rani, D. P., Suresh, D., Kapula, P. R., Akram, C. M., Hemalatha, N., & Soni, P. K. (2021). IoT based smart solar energy monitoring systems. *Materials Today: Proceedings*.

[15] Vhanmane, A. R., & Papade, C. V. (2021). IOT based solar applications. *AIJR Abstracts*, 108.

[16] Zhou, H., Liu, Q., Yan, K., & Du, Y. (2021). Deep learning enhanced solar energy forecasting with AI-Driven IoT. *Wireless Communications and Mobile Computing*, 2021, 9249387.

[17] Ashwin, M., Alqahtani, A. S., & Mubarakali, A. (2021). IoT-based intelligent route selection of wastage segregation for smart cities using solar energy. *Sustainable Energy Technologies and Assessments*, 46, 101281.

[18] Yakut, M., & Erturk, N. B. (2022). An IoT-based approach for optimal relative positioning of solar panel arrays during backtracking. *Computer Standards & Interfaces*, 80, 103568.

[19] Hoque, E., Saha, D. K., & Rakshit, D. (2021). 9 IoT-based intelligent solar energy-harvesting technique with improved efficiency. *Artificial Intelligence and Internet of Things for Renewable Energy Systems*, 12, 279.

[20] Rawat, D. S., & Padmanabh, K. (2021). Prediction of solar power in an IoT-enabled solar system in an academic campus of India. In *Intelligent Systems, Technologies and Applications* (pp. 419–431). Singapore: Springer.

[21] Ram, S. K., Sahoo, S. R., Das, B. B., Mahapatra, K., & Mohanty, S. P. (2021). Eternal-Thing 2.0: Analog-Trojan resilient ripple-less solar energy harvesting system for sustainable IoT in smart cities and smart villages. arXiv preprint arXiv:2103.05615.

[22] Mahadevaswamy, K. K. H., Chethan, K., & Sudheesh, K. V. (2021). Voice controlled IoT based grass cutter powered by solar energy. In *Advances in VLSI, Signal Processing, Power Electronics, IoT, Communication and Embedded Systems: Select Proceedings of VSPICE 2020*, 752, 327.

[23] Ramu, S. K., Irudayaraj, G. C. R., & Elango, R. (2021). An IoT-based smart monitoring scheme for solar PV applications. In *Electrical and Electronic Devices, Circuits, and Materials: Technological Challenges and Solutions*, 211–233.

[24] Ali, M. I., Uthman, M., Javed, S., Aziz, F., & Ali, F. A novel architecture for solar plant monitoring using IoT. In *Intelligent Transportation Systems* (Vol. 19, p. 23).

[25] IoT Based Solar Energy Monitoring System - ScienceDirect. (2021, July 28). IoT based solar energy monitoring system - ScienceDirect. https://www. sciencedirect.com/science/article/pii/S2214785321052238

[26] Gör, İ. (2014). Vektör nicemleme için geometrik bir öğrenme algoritmasının tasarımı ve uygulaması (Master's thesis, Adnan Menderes Üniversitesi).

[27] Amazon Kinesis Data Streams FAQs | Amazon Web Services. (n.d.). Amazon Web Services, Inc. https://www.amazonaws.cn/en/kinesis/data-streams/faqs/

[28] Amazon Kinesis Data Firehose | Amazon Web Services. (n.d.). Amazon Web Services, Inc. https://www.amazonaws.cn/en/kinesis/data-firehose/

[29] Amazon Simple Storage Service (S3) — Cloud Storage — AWS. (n.d.). Amazon Web Services, Inc. https://aws.amazon.com/s3/faqs/?nc=sn&loc=7

[30] Amazon Athena FAQs – Serverless Interactive Query Service – Amazon Web Services. (n.d.). Amazon Web Services, Inc. https://aws.amazon.com/athena/ faqs/?nc=sn&loc=6

[31] Amazon QuickSight - Business Intelligence Service - Amazon Web Services. (n.d.). Amazon Web Services, Inc. https://aws.amazon.com/quicksight/

[32] Serverless Computing - AWS Lambda - Amazon Web Services. (n.d.). Amazon Web Services, Inc. https://aws.amazon.com/lambda/

[33] Amazon SQS | Message Queuing Service | AWS. (n.d.). Amazon Web Services, Inc. https://aws.amazon.com/sqs/

[34] Dogan, A., & Birant, D. (2021). Machine learning and data mining in manufacturing. *Expert Systems with Applications*, 166, 114060.

[35] Zhong, S., Zhang, K., Bagheri, M., Burken, J. G., Gu, A., Li, B., ... & Zhang, H. (2021). Machine learning: new ideas and tools in environmental science and engineering. *Environmental Science & Technology*, 55(19), 12741–12754.

[36] Huang, H. Y., Broughton, M., Mohseni, M., Babbush, R., Boixo, S., Neven, H., & McClean, J. R. (2021). Power of data in quantum machine learning. *Nature Communications*, 12(1), 1–9.

[37] Paullada, A., Raji, I. D., Bender, E. M., Denton, E., & Hanna, A. (2021). Data and its (dis) contents: A survey of dataset development and use in machine learning research. *Patterns*, 2(11), 100336.

[38] Liu, B., Ding, M., Shaham, S., Rahayu, W., Farokhi, F., & Lin, Z. (2021). When machine learning meets privacy: A survey and outlook. *ACM Computing Surveys (CSUR)*, 54(2), 1–36.

[39] Thiebes, S., Lins, S., & Sunyaev, A. (2021). Trustworthy artificial intelligence. *Electronic Markets*, 31(2), 447–464.

[40] Wu, H., & Dai, Q. (2021). Artificial intelligence accelerated by light.

[41] Dhanaraj, R. K., Rajkumar, K., & Hariharan, U. (2020). Enterprise IoT modeling: supervised, unsupervised, and reinforcement learning. In *Business Intelligence for Enterprise Internet of Things* (pp. 55–79). Cham: Springer.

Chapter 2

Study of conventional & non-conventional SEPIC converter in solar photovoltaic system using Proteus

Parul Dubey and Arvind Kumar Tiwari
C.V. Raman University, Chhattisgarh, India

CONTENTS

2.1 INTRODUCTION

In this chapter, the open and closed-loop SEPIC converters are compared, including environmental impact on PV Cell Production of solar photovoltaic harvesting system described by analysis and performance using DC–DC converter, like the SEPIC converter. Photovoltaic system convergence time is reduced using the maximum power-point tracking (MPPT) algorithm [1], responding faster to atmospheric changes than a conventional algorithm with minimum ripple content in the output. A digital controller is used to control the DC–DC SEPIC power converter. As a result, renewable energy sources have significant development potential, and natural gas and renewable energy have been the primary sources of increased capacity in recent years. Solar and wind power account for most of the growth in renewable energy. Many reasons, including renewable portfolio criteria, lowering installation prices, and incentives, have contributed to the world's consistent and rapid rise in solar photovoltaic installation [2]. The rising expansion of solar PV capacity and the construction of larger power plants have stimulated research and development in high-power converter topologies for PV applications [3]. Since solar energy is the raw material and the primary energy source for various renewable energy applications, understanding the intensity of solar irradiation is critical for system performance. Electric energy sources that can fulfill society's expanding demands while having a common environmental effect and great efficiency have been investigated in recent years. In this context, photovoltaic cells convert electrical energy and become one of the world's most well-known and commonly used resources. On the other hand, the existence of clouds in the sky is the most unexpected element impeding the capture of solar irradiation and the correct conversion of sunlight into power [4]. The design of a solar inverter that effectively converts DC power to AC power is a multi-domain problem [5]. This entails current and voltage detection, driving the measured power, power-point tracking, DC–DC conversion, DC-AC inversion, grid protection, and so on. It is recommended that these internal work modules be designed with great efficiency to build an efficient solar inverter. To accomplish these tasks quickly, with low energy consumption, and minimal power loss, an example architecture includes field-programmable gate arrays (FPGAs) that are linked to send the generated electricity to the grid [5]. A DC–DC converter sends data to some PV panels, where it is processed. Power and voltage are increased in this converter's outputs to meet grid feeding needs. The solar panels' power production is also monitored and adjusted using sensors such as those used to measure current and voltage [5]. This control is accomplished by using algorithms that track the maximum power-point of PVs to maximize their output. DC–DC controllers, which use low-power drivers, help in this effort. PV inverters, or DC-AC converters, convert DC voltages to AC voltages for grid feeding. To begin, a DC–DC converter sends data to several PV panels. This converter increases output power and voltage to meet grid feeding needs. The solar panels' output power is monitored and adjusted via current and voltage sensors. Because MPPT algorithms strive to optimize PV attributes for maximum output, this control can be achieved. They used low-power drivers, such as those found in DC–DC controllers, to aid this effort. Solar PV inverters use measured power to convert DC voltage to AC voltage for grid feeding. As soon as the grid goes down, the feeding interface cuts off solar electricity from the rest of the system [5]. Numerous methods can be used to design PV inverter circuits, each with a unique goal. In the next section, we shall look at how well these algorithms perform under various electrical conditions. After that, a statistical examination of multiple systems will help researchers uncover the best inverter designs for their PV systems. It concludes with several intriguing conclusions concerning these models and suggestions for further improvement [5, 6].

Figure 2.1 shows the SEPIC model for high stability. Many algorithms are available for developing PV inverter circuits, which differ in efficiency [5], application, power needs, etc. The next section will examine these algorithms to see if they suit various electrical conditions. A statistical study of multiple systems is performed to help researchers determine the optimal inverter design for their PV system. An array of PV inverter circuit methods are accessible, each with unique features such as conversion efficiency, application, and power demands. These algorithms will be examined in detail in the next part to determine their effectiveness in various electrical scenarios. Thus, the best practicable inverter designs for each unique PV system design are determined by conducting a statistical study of multiple systems.

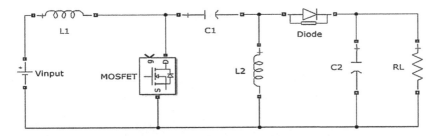

Figure 2.1 The SEPIC model for high stability [5].

1) Performance
   An irregular heart rate is used to swiftly switch on and off a contaminated MOSFET in all DC–DC converters. What the interpreter does as a consequence is what distinguishes SEPIC. To charge inductor 1 to its maximum voltage and charge inductor 2 via capacitor 1, a MOSFET in SEPIC is turned on when the signal is too strong. To store the yield, diode 2 has been turned off. Capacitors and inductors are charged when the power is low, or the MOSFET has been turned off, respectively [7]. Much time (movement cycle) is spent driving down. Therefore, the effect may be much more noticeable. This is because the more time the examiners take, the more intense their pressure will be [7]. However, if the beat continues for an excessive amount of time, the capacitors will not be able to charge, and the converter will malfunction [8].

2) Specs
   Particular requirements dictate that the converter must fulfill a voltage range of 6 V to 18 V and a current draw of 1 A. The output coefficient is 0.5 A, and the maximum output power is 0.1 V.

3) Work cycle calculation
   The number of SEPIC converters in use changes with the duty cycle and local pesticides [7]. When a decent SEPIC converter is released, [8]

$$V_o = (D * V_{IN}) / (1 - D) \qquad (2.1)$$

But this does not address losses caused by pesticides such as diode drop V_d, which is widely used (0.5 V). As a result, the computation is made,

$$V_o + V_d = (D \times V_{IN}) / (1 - D) \qquad (2.2)$$

This is,

$$D = (V_o + V_d) / (V_{IN} + V_o + V_d) \qquad (2.3)$$

Table 2.1 Variables for the SEPIC converter that have been designed

Specification of the SEPIC converter's variables	
Content	Values
Switching frequency [7], f_{sw}	17.5 KHz
MOSFET [7]	IRF840
Inductor, L1, L2 [7]	600 μH, 600 μH [7]
Duty Cycle, D [7]	40.98 to 67.56% [7]
Capacitor, C1, C2, C3 [7]	470 μF, 100 μF, 100 μF [7]

When V_{IN} is at its lowest point, the duty cycle will be at its highest point, and when V_{IN} is at its highest point, it will be at its lowest point [7].

4) Calculation of an inductor

Theoretically, as inductors get larger, the area will begin to operate, and the likelihood of an explosion will decrease [7]. Larger inductors, on the other hand, are more expensive, and have a greater degree of internal blockage. This excellent inner impediment will aid the performance of the converter positively. It is necessary to pick four inductors that are adequate to keep the voltage and ripple current at an appropriate level to construct a leading converter [8]. The intended SEPIC converter parameters derived from the preceding equations are displayed in Table 2.1 shows the Designed Parameters for SEPIC Converter.

$$L = \left(V_{in\,min} + D_{max}\right) / \left(I_{ripple\,max} \times f_{sw}\right) \tag{10}$$

2.2 SEPIC CONVERTER IN THE OPEN-LOOP SYSTEM

In a type of converter known as a Single-Ended Primary Inductance Converter [7], the voltage output can be greater or less than the voltage applied to it (SEPIC) [9]. In reality, the Boost and Buck converters coexist in this architecture. The circuit will function as a buck converter if the duty factor is less than 50% [10]. A boost converter is activated when the duty factor exceeds 50%. A boost converter can be used if the duty factor is more than 50% [10]. The inductors power the converter in continuous mode to maintain a constant current. Converter enters discontinuous mode by decreasing the inductor value [11].

The voltage output may be approximated as: $V_{OUT} = (V_{IN} \times D) / (1 - D)$, where:

D = T1/T (duty factor)

T = PWM period

T1 = OFF time

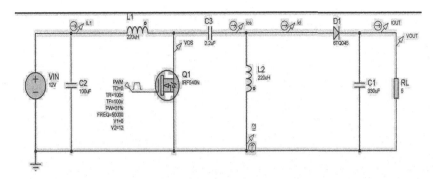

Figure 2.2 Open-loop SEPIC converter.

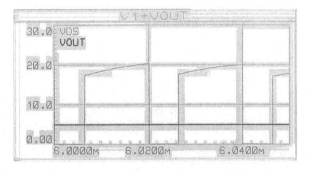

Figure 2.3 Waveform of V_{DS} and V_{OUT} of SEPIC converter.

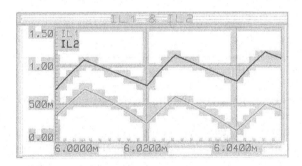

Figure 2.4 Waveform of IL1 and IL2 of SEPIC converter.

In the T = 20 us and T1 = 6.3 us (D = 31%) example, V_{OUT} is about 5 V, and the circuit behaves like a buck converter [12].

Table 2.2 shows the reading of different values of input voltages, from 06 v to 18 v, with additional duty cycles. There is a boost converter if the duty cycle is larger than 50% and a buck converter if the duty cycle is less than 50%. Different input voltages have different duty cycles, so their operation is based on the duty cycle as it performs boost or buck. Input voltage 6 to

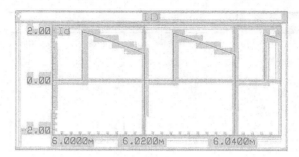

Figure 2.5 Waveform of I$_D$ of SEPIC converter.

Table 2.2 Observation table of open-loop SEPIC converter

Ser	V$_{IN}$	Duty cycle (%)	V$_{OUT}$	I$_{OUT}$
I	06	68.4	2.34	0.46
2	07	62.60	2.8029	0.50
3	08	59.25	3.26	0.65
4	09	54.36	3.74	0.74
5	10	52.35	4.18	0.83
6	11	51.14	4.64	0.92
7	12	34.87	5.0	1.01
8	13	27.28	5.55	1.11
9	14	25.11	5.99	1.19
10	15	23.89	6.44	1.12
11	16	21.30	6.91	1.30
12	17	16.55	7.38	1.42
13	18	11.09	7.84	1.56

11 have a more than 50% duty cycle, so they work as a boost converter. Input voltages 12 to 18 have a duty cycle of less than 50%, so they work as buck converters.

When the input voltage is 6 V, its duty cycle is 68.4. So it works as a boost converter at that time. Its V_{OUT} and I_{OUT} are 2.34 and 0.46 resp. Figure 2.6 shows the V_{OUT} and I_{OUT} response waveform when V_{IN} is 6v.

When the input voltage is 7 V, its duty cycle is 62.60%. So it works as a boost converter. At that time, its V_{OUT} and I_{OUT} are 2.8029 and 0.50 resp. Figure 2.7 shows the V_{OUT} and I_{OUT} response waveform when V_{IN} is 7 v.

When the input voltage is 9 v, its duty cycle is 54.36%. So it works as a boost converter at that time. Its V_{OUT} and I_{OUT} are 3.74 and 0.74 resp. Figure 2.8 shows the V_{OUT} and I_{OUT} response waveform when V_{IN} is 9v.

When the input voltage is 14 V, its duty cycle is 25.11%. So it works as a buck converter. At that time, V_{OUT} and I_{OUT} are 5.99 and 1.19 resp. Figure 2.9 shows the V_{OUT} and I_{OUT} response waveform when V_{IN} is 14v.

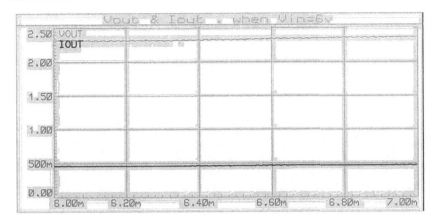

Figure 2.6 Waveform of V_{OUT} and I_{OUT} at V_{IN} = 6 V.

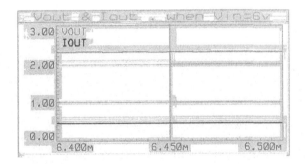

Figure 2.7 Waveform of V_{OUT} & I_{OUT} at V_{IN} = 7 V.

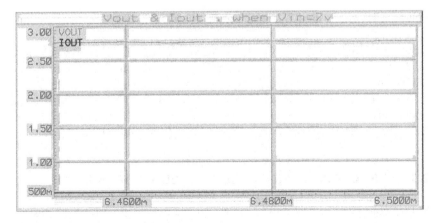

Figure 2.8 Waveform of V_{OUT} & I_{OUT} at V_{IN} = 9 V.

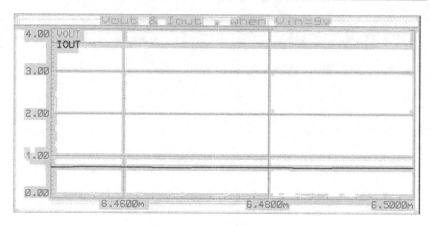

Figure 2.9 Waveform of V_{OUT} & I_{OUT} at V_{IN} = 14 V.

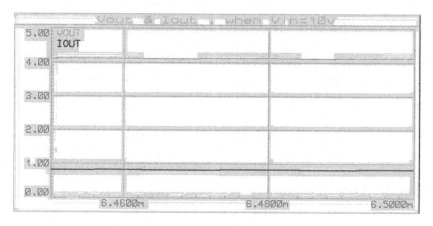

Figure 2.10 Waveform of V_{OUT} & I_{OUT} at V_{IN} = 16 V.

When the input voltage is 16 V, its duty cycle is 21.30%. So it works as a buck converter. At that time, V_{OUT} and I_{OUT} are 6.91 and 1.30 resp. Figure 2.10 shows the V_{OUT} and I_{OUT} response waveform when V_{IN} is 16v.

When the input voltage is 17 V, its duty cycle is 16.55%. So it works as a buck converter. At that time, V_{OUT} and I_{OUT} are 7.38 and 1.42 resp. Figure 2.11 shows the V_{OUT} and I_{OUT} response waveform when V_{IN} is 17 V.

2.3 SEPIC CONVERTER IN THE CLOSED-LOOP SYSTEM

SEPIC converter in a closed-loop system involves maintaining a constant output voltage irrespective of supply or load using the automatic SEPIC converter and microcontroller (LPC2148). The SEPIC converter with LPC2148

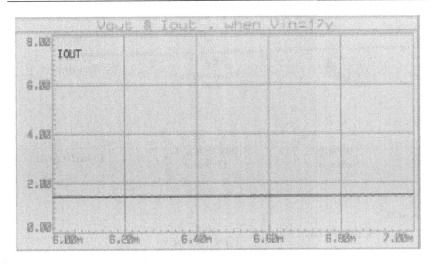

Figure 2.11 Waveform of V_{OUT} & I_{OUT} at V_{IN} = 17 V.

is intended to work with suitable loads to extract the most power from solar modules. The photovoltaic (PV) module feds output to DC–DC SEPIC as well as to the voltage sensing circuit in the feedback path; the other comparator circuit compares the sense voltage with the set point and gives the difference signal to Peripheral Interface Controller (PIC) microcontroller, the photovoltaic (PV) module creates the direct current (DC) voltage that corresponds to sunshine radiation. Changes in atmospheric circumstances develop variations in the amount of sunlight landing on the PV module, resulting in significant variance in output. In ideal conditions, the PV module produces a maximum output voltage of roughly 20.6 V, but variations in sunshine exposure induce a voltage decrease from 20.6 V to 6 V. The DC–DC SEPIC converter uses this variable voltage to adjust the output voltage, which may be modified from 24 V to 6 V. The feedback circuit uses a voltage divider network to return the observed voltage to the comparator. The comparator compares the detected output to the set point voltage and produces an error voltage. The erroneous voltage is then sent into the PIC microcontroller's analog to a digital converter (ADC) [13]. The PIC microcontroller uses Pulse Width Modulation to create pulses matching the error signal (PWM). Finally, these pulses govern the switching processes of the DC–DC SEPIC converter [7], allowing the output voltage to be kept constant by altering the width of the PWM pulses. This constant voltage may be obtained for various applications, including dc loads, battery charging, and DC/AC grids [13].

The SEPIC converter with LPC2148 is intended to work with suitable loads to get the most power out of solar modules. The purpose of a SEPIC

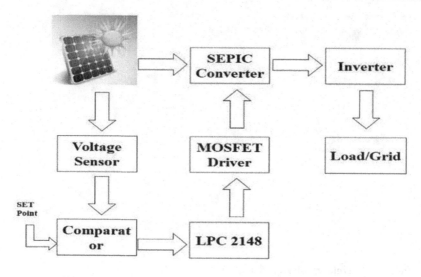

Figure 2.12 The suggested system's block diagram.

converter with a microcontroller is to keep the output voltage constant. Figure 2.12 depicts the Proteus circuit diagram and the closed-loop SEPIC converter idea. There are more software and hardware options in Proteus VSM than in microcontroller programming [13]. Academics and engineers widely use the tool. The controller optimizes a solar-cell module's output power, which manages its duty cycle to ensure that the solar-cell module is always operating at maximum output power. In response to the output voltage of the PV module, the computer changes the duty cycle of the conversion system. When the ultimate power-point loop is activated, the reference voltage is established, and the voltage regulator loop regulates the solar output voltage by the reference voltage setting [13].

A wide range of products and equipment is included. Many analysts and designers use the Proteus for quick and easy testing [7, 14] simulation. The output power of photovoltaic modules increases as the activity cycle changes, maintaining maximum efficiency [7, 14]. The microcontroller performs automatic addition [14]. Or by the PV module's output power decreasing the converter's activity cycle [7]. The standard recommends that the day-light-based yield is manipulated using the voltage controller circle. And set as the best operating point and the control device to establish the VREF evaluation input is illustrated in Figure 2.14. The LPC micro-ADC controller uses the VO aspect to create the error energy. The voltage delivered to the non-switching comparator is a sensing element in a detachable network. The measurement's output is converted by the LPC microcontroller's ADC [7]. There are various ways to convey the idea of more authority [7]. In addition to the flowcharts, the gadget has been optimized for use of the go [14].

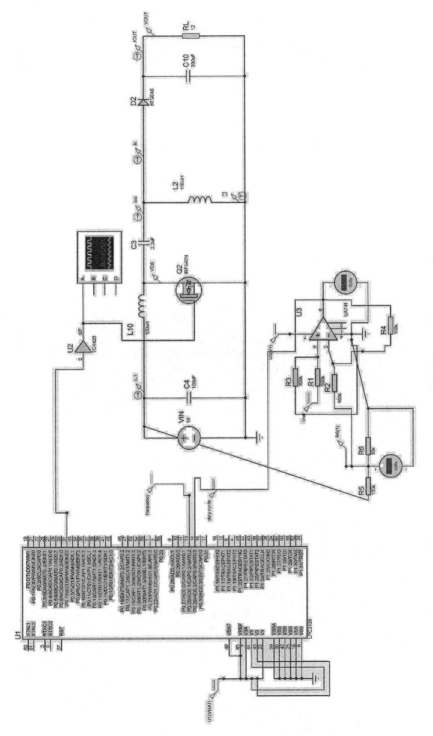

Figure 2.13 Complete circuit diagram of SEPIC converter with MPPT using Proteus software.

Table 2.3 Values of a passive component in SEPIC converter

Ser	Component	Rating
1	C_2	100 μF
2	L_1	600 μH
3	L_2	600 μH
4	C_3	100 μF
5	C_1	470 μF

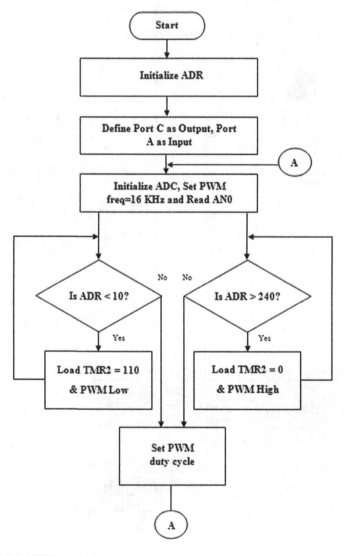

Figure 2.14 MPPT algorithm.

Table 2.4 The SEPIC converter utilizes a microprocessor to maintain a constant output voltage

Parameters of solar PV modules			
V_{IN} (V) [7]	Duty cycle (%) [7]	V_{OUT} (V)	I out (A)
18	49.41	25.2	1.1
16	55.29	25	1.1
14	60	24.8	1.1
12	65.38	24	1.03
10	71.76	23.9	1.0
8	75.29	23.7	1.0
6	80	23	1.0

Based on Proteus, the simulation came out right: SEPIC and LPC2148 are to be used in parallel to maximize power from the PV panel [7, 14]. The converter topology has a microprocessor that ensures the output voltage stays the same [7].

Table 2.4 demonstrates the reading of various input voltages at different times. With a range of voltages from 18 V to 0.6 V and a variety of duty cycles. It is a boost converter if the duty cycle exceeds 50%; otherwise, it is considered a buck converter [15]. The system is classified as a buck converter if the duty cycle is less than 50%. Different input voltages have different duty cycles, so their operation is based on the duty cycle as they perform boost or buck. Input voltages 0.6 to 11 have a more than 50% duty cycle, so they work as a boost converter. Input voltages 12 to 18 have a duty cycle of less than 50%, so they work as a buck converter [16].

Figures 2.15 and 2.16 are the Proteus simulation results, showing that V_{IN} adjustment maintains a consistent 24 V yield voltage and I_{OUT} is approximately 1A [7, 14] by red and green color, respectively.

Proteus simulations show that the resulting pulse's duty cycle is [7] modified (49%–80%) in response to variations in solar panel output to capture the maximum power-point (Figures 2.17 and 2.18)

SPWM signals are created using the PIC16F877A microcontroller to operate the MOSFET-based inverter, as shown in the flowchart in Figure 2.20. Proteus' MOSFET-based H-bridge inverter output is shown in Figure 2.21.

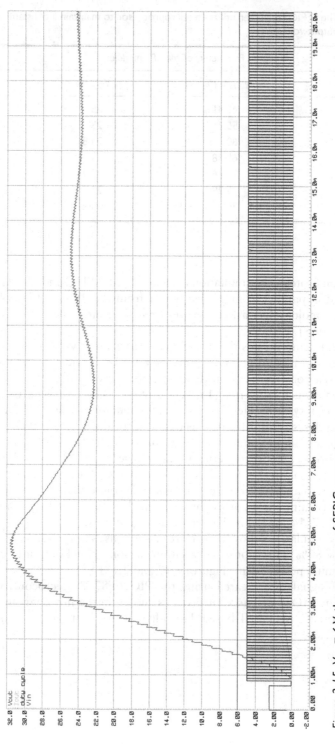

Figure 2.15 V_{IN} = 6 V, the response of SEPIC converter.

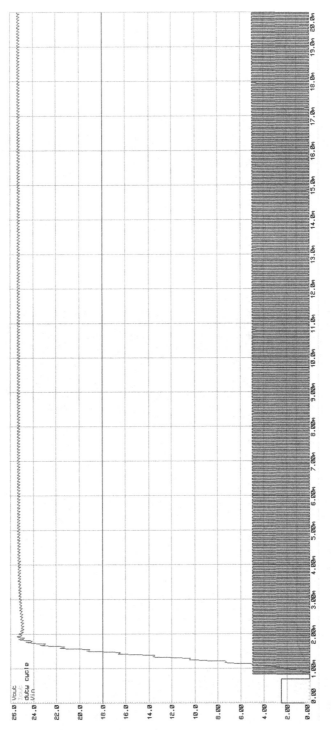

Figure 2.16 SEPIC converter response when V_{IN} is 18 V.

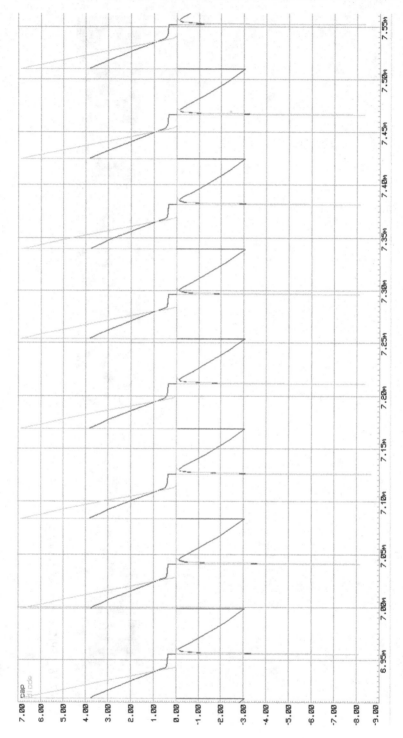

Figure 2.17 Capacitor charge/discharge waveforms and diode switching [7].

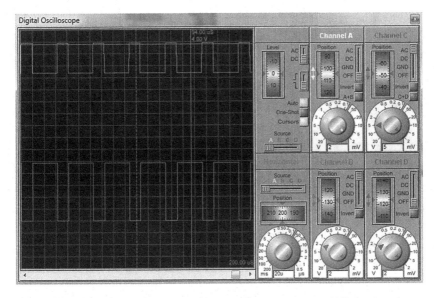

Figure 2.18 Proteus Simulation developed PWM pulses with 49% duty cycle.

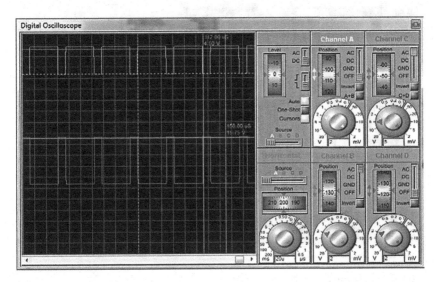

Figure 2.19 Proteus Simulation developed PWM pulses with 80% duty cycle.

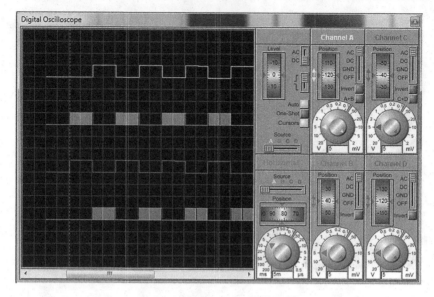

Figure 2.20 Proteus simulation of SPWM pulses for inverters.

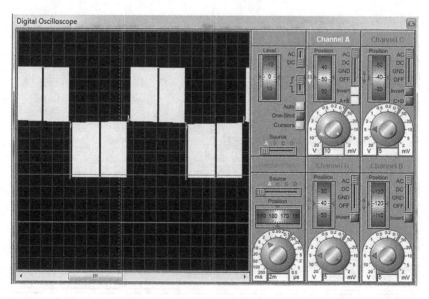

Figure 2.21 Proteus inverter output.

2.4 CONCLUSION

This project aims to keep the PV operating voltage around maximum for fluctuating climatic circumstances by modifying the maximum power operating point (MPOP) [17]. In the first observation, we observed an open-loop SEPIC converter. Its V_{OUT} and I_{OUT} waveforms with different input voltages and duty cycles. In the second observation, we observed a closed-loop SEPIC converter. It is a V_{OUT} and I_{OUT} waveform with varying input voltage and duty cycles [18]. A digitally controlled SEPIC converter for solar PV with voltage and current sensing is presented [19]. Maximum powerpoint is estimated by sensing voltage and current, which varies the pulse's duty cycle and frequency, respectively, to minimize the power loss caused by switching losses and improve the system's overall performance. The comprehensive observation shows that if the input voltage of the PV module has been changed due to environmental changes, there are no changes in V_{OUT} and I_{OUT} because of the closed-loop SEPIC converter.

REFERENCES

[1] Anil Kumar Saini and Ashish Kumar Dubey, "Performance Analysis of Single Phase Induction Motor with Solar PV Array for water Pumping System," *Int. J. Eng. Res.*, vol. V6, no. 04, 2017. doi: 10.17577/ijertv6is040567.

[2] D. Gielen, F. Boshell, D. Saygin, M. D. Bazilian, N. Wagner, and R. Gorini, "The Role of Renewable Energy in the Global Energy Transformation," *Energy Strategy. Rev.*, vol. 24, no. June 2018, pp. 38–50, 2019. doi: 10.1016/j.esr.2019.01.006.

[3] E. Rakhshani, K. Rouzbehi, A. J. Sánchez, A. C. Tobar, and E. Pouresmaeil, "Integration of Large Scale PV-based Generation into Power Systems: A Survey," *Energies*, vol. 12, no. 8, 2019. doi: 10.3390/en12081425.

[4] H. J. Loschi, Y. Iano, J. León, A. Moretti, F. D. Conte, and H. Braga, "A Review on Photovoltaic Systems: Mechanisms and Methods for Irradiation Tracking and Prediction," *Smart Grid Renew. Energy*, vol. 06, no. 07, pp. 187–208, 2015. doi: 10.4236/sgre.2015.67017.

[5] S. R. Hole and A. D. Goswami, "Quantitative Analysis of DC–DC Converter Models: A Statistical Perspective Based on Solar Photovoltaic Power Storage," *Energy Harvest. Syst.*, vol. 9, no. 1, pp. 113–121, 2022. doi: 10.1515/ehs-2021-0027.

[6] P. L. R. Bharat and S. Dhak, "Evaluation of Kernel-Level IoT Security and QoS Aware Models from an Empirical Perspective," *Springer Sci. Bus. Media LLC, 2022*, vol. 851, no. 1–2, pp. 731–746, 2018. doi: 10.1007/s35152-018-0008-5.

[7] A. Das Hole Shreyas Rajendra and Goswami, "Maintain Maximum Power Point Tracking of Photovoltaic using SEPIC Converter," in *2022 2nd International Conference on Power Electronics & IoT Applications in Renewable Energy and its Control (PARC)*, 2022, pp. 1–6. doi: 10.1109/parc52418.2022.9726607.

[8] G. Iranian Journal of Electrical and Electronic Engineering, "Sepic Converter Design and Operation," *Submitt. 5 / 1 / 14 Partial Complet. Requir. a By Advis. Alex Emanuel*, [Online]. Available: http://docplayer.net/31187215-Sepic-converter-design-and-operation.html

[9] U. Ajith Kumar, M. Boobana, M. Ramyaa, A. Thangam, R. Abinaya, and V. Seenivasan, "Designing of Single Ended Primary Inductance Converter for Solar PV Application Using Arduino Controller," *Int. Res. J. Eng. Technol.*, vol. 5, no. 4, pp. 1556–1560, 2018.

[10] I. D. Jitaru and S. Birca-Galateanu, "Small-signal Characterization of the Forward-Flyback Converters with Active Clamp," *Conf. Proc. - IEEE Appl. Power Electron. Conf. Expo. - APEC*, vol. 2, no. 520, pp. 626–632, 1998. doi: 10.1109/apec.1998.653965.

[11] K. L. Barui and B. Bhattacharya, "Development of Non-destructive Quality Assessment Criterion of 7075-T6 Al-Zn-Mg Alloy Components Using Fracture Mechanics Techniques," *Eng. Fract. Mech.*, vol. 47, no. 1, pp. 121–131, 1994. doi: 10.1016/0013-7944(94)90242-9.

[12] S. L. Tripathi, P. A. Alvi, and U. Subramaniam, *Electrical and Electronic Devices, Circuits and Materials: Design and Applications* (1st ed.). CRC Press, 2021. doi: https://doi.org/10.1201/9781003097723.

[13] R. Kansal and M. Singh, "PIC Based Automatic Solar Radiation Tracker," *Asian J. Chem.*, vol. 21, no. 10, 2009.

[14] A. V. Padgavhankar and S. W. Mohod, "Experimental Learning of Digital Power Controller for Photovoltaic Module Using Proteus VSM," *J. Sol. Energy*, vol. 2014, pp. 1–8, 2014. doi: 10.1155/2014/736273.

[15] S. P. Suman Lata Tripathi, Mithilesh Kumar Dubey, and Vinay Rishiwal, *Introduction to AI techniques for Renewable Energy System*. 2021. doi: https://doi.org/10.1201/9781003104445.

[16] M. D. Cookson and P. M. R. Stirk, *Green Energy: Fundamentals, Concepts, and Application*. 2019. doi: 10.1002/9781119760801

[17] K. M. Bataineh and A. Hamzeh, "Efficient Maximum Power Point Tracking Algorithm for PV Application under Rapid Changing Weather Condition," *ISRN Renew. Energy*, vol. 2014, pp. 1–13, 2014. doi: 10.1155/2014/673840.

[18] S. L. Tripathi, P. A. Alvi, and U. Subramaniam, *Electrical and Electronic Devices, Circuits and Materials: Technological Challenges and Solutions*. doi: 10.1002/9781119755104.

[19] C. L. Kumari, V. K. Kamboj, S. K. Bath, S. L. Tripathi, M. Khatri, and S. Sehgal, *A Boosted Chimp Optimizer for Numerical and Engineering Design Optimization Challenges*, no. 0123456789. Springer, London, 2022. doi: 10.1007/s00366-021-01591-5.

Chapter 3

Design and implementation of non-inverting buck converter based on performance analysis scheme

M. Siva Ramkumar and R. Felshiya Rajakumari
Karpagam Academy of Higher Education, Coimbatore, Tamil Nadu, India

CONTENTS

3.1 INTRODUCTION

Many surveyors lecture the evolution of converters in different schemes for industrial purposes [1, 2, 3]. Since on the whole part of the listener loads and cargo space essentials use DC supply, DC–DC schemes have purchase reputation around the surveying area. In this tabloid, non-inverting buck-boost converter operates in dual operation. It has more numeral of cargo space devices and it is broadly used in succession equipment [4, 5]. The idea of this broadsheet is to recommend a non-inverting buck-boost converter for productivity voltage ripple reduction procedures [6, 7]. Simulation studies of the topologies are carried out in MATLAB/Simulink. Sector I discuss the Introduction of the work. Portion II describes the non-reversing buck-boost converter. Sector III states the simulation results of NIBB converter. Sector IV imitates the computation of output voltage ripple for NIBB converter based on different duty cycle values. Portion V intimates the conclusion of the work.

3.2 NON-REVERSING BUCK-BOOST CONVERTER

The non-reversing topology harvests a yield voltage that is of the identical divergence as the effort voltage. In the buck mode, the amount produced voltage is unwavering by the maneuver of the MOSFET and diode D1. In the boost mode, the productivity voltage is resolute by the operation of the MOSFET2 and diode D2. Depending on the ratio between the idea voltage and the harvest voltage, the non-reversing buck-boost converter can activate in buck mode or boost mode. In order to increase the efficiency, the operating modes are: converter operates both as buck converter (only switch T1 is measured, while T2 is always turned off) or boost converter (only switch T2 is precise, while T1 is always curved on) [3, 5, 8, 9]. Different control procedures can be pragmatic to the non-inverting buck-boost converter, depending on the application. Figure 3.1 confirms the circuit plan for non-inverting buck-boost converter.

3.2.1 Operating modes of NIBB converter-buck mode

Figure 3.2 confirmations the correspondent circuit of NIBB converter for buck operation. It controls in two modes:

Mode 1: Figure 3.3 appearances the mode 1 correspondent circuit for NIBB converter based on buck operation. During approach 1, the swapping transistor Qs is substituted on. Since the key in voltage is grander than the productivity voltage, the current in the inductor increases linearly during this interval [8].

Mode 2: Figure 3.4 shows the mode 2 correspondent circuit for NIBB converter based on buck operation. During mode 2 operation swapping transistor QS is switched off. Freewheeling diode gets comportment in mode 2 operation because it is in frontward bias just as the inductor voltage antitheses its polarity. Here, the inductor current sprays as the energy stored in it transferred to the capacitor and expanded by the load.

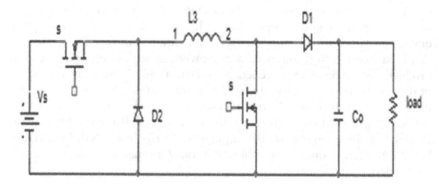

Figure 3.1 Circuit diagram of NIBB converter.

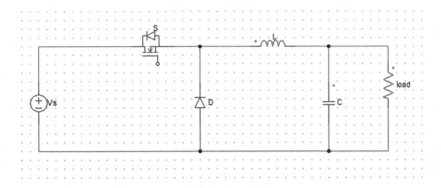

Figure 3.2 Correspondent circuit for NIBB converter based on Buck converter operation.

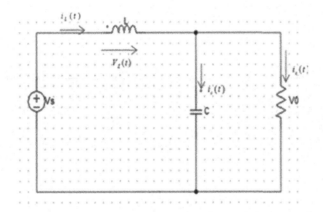

Figure 3.3 Mode 1: Correspondent circuit for NIBB converter based on Buck converter operation.

3.2.2 Operating modes of NIBB converter – boost mode

Figure 3.5 Correspondent circuit of NIBB converter for Boost operation. It activates in two styles of operations [8, 10].

Style 1: Figure 3.6 Mode 1: equivalent circuit for NIBB converter based on boost operation.

During on-time, button(s) is on at $t = 0$ and it sacks at $t =$ ton. Freewheeling diode is converse bias since the power drop across the collector-emitter junction of switch is slighter than the productivity voltage. During on-time, the inductor current rises linearly. The production current during this interval is delivered entirely from the production capacitor, which is to provide sufficient load current during the on-time, with a slight, precise drop in output current [10, 11].

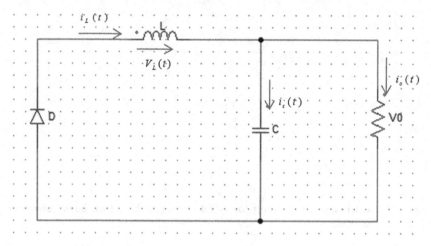

Figure 3.4 Mode 2: Correspondent circuit for NIBB converter based on Buck converter operation.

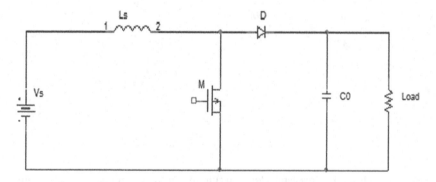

Figure 3.5 Correspondent circuit for NIBB converter based on Boost converter operation.

Style 2: Figure 3.7 expressions the mode 2 correspondent circuits for NIBB converter based on Boost operation.

During off-time, button is off at $t = t_{on}$, since the current in the inductor cannot variation simultaneously, the voltage across the inductor converses its division to sustain a constant current. Now the current will stream through the L, C, and load [12, 13]. The inductor conveys its stowed energy to the output capacitor and charges to the FWD to an elevated voltage than the effort voltage. During off-time, the inductor current cascades linearly. The energy provisions the current and restocks the charge worn out away from the productivity capacitor, when it was unaided supplying the load current during the prompt [14, 9].

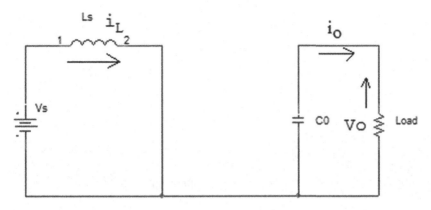

Figure 3.6 Style 1: Correspondent circuit for NIBB converter based on Boost converter operation.

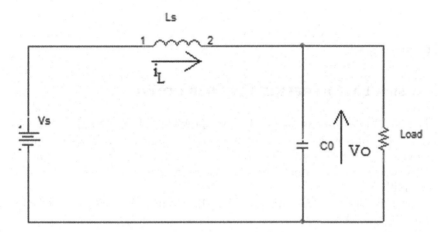

Figure 3.7 Mode 2 correspondent circuit for NIBB converter based on Boost converter operation.

3.2.3 Design equations of NIBB converter [15]

Conversion Gain:

The alteration gain for the boost converter is given in Equation (3.1)

$$V_0 = \frac{V_s}{1-D} \tag{3.1}$$

where, V_s the foundation voltage, D is the duty cycle for the boost converter, V_0 is the production voltage.

Inductance:
The peak-to-peak inductor current ripple is given as in Equation (3.2)

$$\Delta I = \frac{V_S D}{f_S L}$$

$$L = \frac{V_S D}{f_S L}$$

(3.2)

where, L is the inductor.
Capacitance:
The peak-to-peak capacitor voltage ripple is given by Equation (3.3)

$$\Delta V_C = \frac{I_a D}{f_S C}$$

$$C = \frac{I_a D}{f_S L}$$

(3.3)

where, f_S is the switching frequency, C is the Capacitor, ΔV_c is the peak-to-peak capacitor

3.3 SIMULATION RESULTS OF NIBB CONVERTER

The NIBB converter is designed in MATLAB/Simulink. The Simulink diagram of NIBB converter is shown in Figure 3.8. Figure 3.9. shows the output voltage of NIBB converter. Figure 3.10 shows the output voltage ripple of NIBB converter. The simulation parameters for NIBB converter are listed in the Table 3.1.

From Figure 3.9 it is contingent that the production voltage of the NIBB Converter is 24 V.

From Figure 3.10 it is contingent the harvest voltage ripple of non-inverting buck-boost converter is 0.03 V.

From Figure 3.11 it is contingent the input current ripple of non-inverting buck-boost converter is 0.35 A.

Table 3.1 Simulation parameters for NIBB converter

Parameters	Values
Source Voltage	12 (volts)
Inductor	40 (µH)
Capacitor	500 (µF)
Resistive Load	100 (ohm)
Switching frequency	100 (kHz)

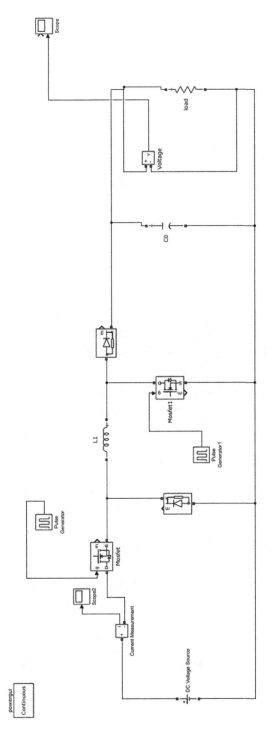

Figure 3.8 Simulink diagram of NIBB converter.

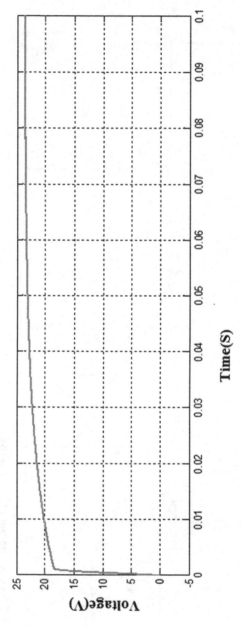

Figure 3.9 Output voltage of NIBB converter.

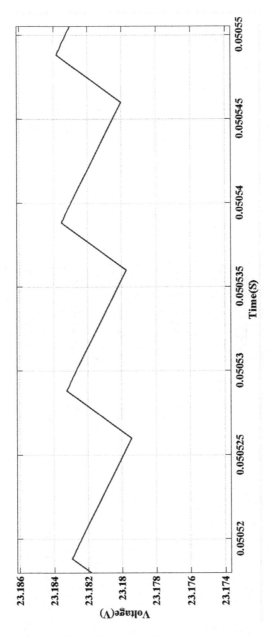

Figure 3.10 Output voltage ripple of NIBB converter.

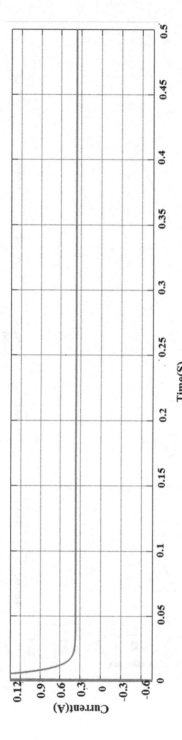

Figure 3.11 Input current ripple of NIBB converter.

3.4 COMPUTATION OF OUTPUT VOLTAGE RIPPLE FOR NIBB CONVERTERS BASED ON DIFFERENT DUTY CYCLE

3.4.1 Values

The recital parameters of NIBB converters are deliberate based on various duty cycles as shown in Table 3.2. Figure 3.12 indicates the duty cycle V_s output voltage ripple of NIBB converters. Based on different duty cycles the NIBB converter performed for future work of electric vehicle applications.

Therefore output voltage ripple and input current ripple found out for NIBB converter.

Table 3.2 Comparison of NIBB DC–DC converters based on different duty cycles

	Parameters	
Duty cycle	Output Voltage Ripple (Volts) & Input Current Ripple (Amps)	
	NIBB Converter	
Cycle values	OVR (Volts)	ICR (Amps)
0.3	0.03	0.35
0.4	0.038	0.42
0.6	0.045	0.50
0.7	0.05	0.58
0.9	0.059	0.65

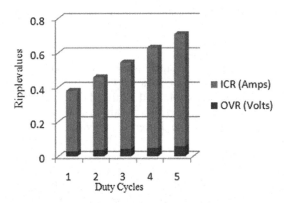

Figure 3.12 Duty cycle V_s output ripple voltage & input current ripple of NIBB converters.

3.5 CONCLUSION

In this paper, the NIBB converter was examined and analyzed in detail. NIBB converter operation operates in boost mode as well as buck mode. Modes of operation and equations were discussed. Further simulation parameters were found for the designing mode. From the mode, constraints were performed, such as production voltage ripple and key in current ripple. The NIBB converters were simulated in MATLAB/Simulink. Comparing to other converter topologies, NIBB converters gives the best performance based in different duty cycles values. Further, this converter can be merged with PV panel for electric vehicle applications.

REFERENCES

[1] Anas Boutaghlaline, Karim el Khadiri, et al., "Design of a Non-Inverting Buck-Boost Converter Controlled by Voltage-Mode PWM in TSMC 180nm CMOS Technology", *Digital Technologies and Applications*, 2021. doi: 10.1007/978-3-030-73882-2_147

[2] Husan Liao, Jiann-Fuh Chen, "A Two Phase Non-inverting Buck Boost Converter for an RL-load", *The Journal of Engineering*, vol. 2022, no. 2, 2021.

[3] Shridhar Sholapur, K.R. Mohan, T.R. Narsimhegowda, "Boost Converter Topology for PV System with Perturb and Observe MPPT Algorithm", *IOSR Journal of Electrical and Electronics Engineering*, vol. 9, no. 4, pp. 50–56.

[4] Srdjan Lale, Milomir Soja, et al., "A Non-Inverting Buck-Boost Converter with an Adaptove Dual Current Mode Control", *FACTA Universitatis Series Electronics and Energetics*, vol. 30, no. 1, January 2017, pp. 67–80.

[5] Ramanjaneya Reddy, Beeramangalla Lakshminarasaiah Narasimharaju, "Single-Stage Electrolytic Capacitor Less Non-Inverting Buck Boost PFC based AC-DC Ripple free LED Driver", *IET Power Electronics*, vol. 10, no. 1, 2017, pp. 38–46.

[6] NIBB Converter- Available: https://in.mathworks.com/help/physmod/ sps/ examples/buck-boost converter

[7] Erik Schaltz, et al., "Non-inverting Buck-Boost Converter for Fuel Cell Applications", *2008 34th Annual Conference of IEEE Industrial Electronics*, 23 January 2009.

[8] G. Bharathi, K. Rajesh, "Stability Analysis of DC–DC Boost Converter for Solar Power Application", *International Journal of Ethics in Engineering and Management Education*, vol. 4, no. 12, 2017. ISSN: 2348-4748.

[9] Chia-Ching Lin, Lung-Sheng Yang, "Study of DC–DC Converters for Solar LED", *IEEE Transactions on Power Electronics*, vol. 13, no. 6, 2012, [no pagination].

[10] Angie Alejandra Rojas Aldana, Oscar Ernesto Beltrain, "Design and implementation of a DC–DC Converter for Photovoltaic Applications", *IEEE Transactions on Power Electronics*, vol. 15, no. 2, 2015, pp. 4673–6605.

[11] Moumita Das, Vivek Agarwal, "A Novel, High Efficiency, High Gain, Front End DC–DC Converter for Low Input Voltage Solar Photovoltaic Applications", *IEEE Transactions on Power Electronics*, vol. 12, no. 2, 2015, [no pagination].

[12] A.W.N. Husna, S.F. Siraj, M.Z. Ab Muin, "Modelling of DC–DC Converter for Solar System Applications", *IEEE Transaction on Power Electronics*, vol. 12, no. 6, 2014, [no pagination].

[13] Zakaria Sabiri, Nadia Machkour, Mamadou Bailo, "DC–DC Converters for Photovoltaic Applications Modelling and Simulations", *IEEE Transactions on Industrial Electronics*, vol. 31, no. 5, 2013, pp. 4799–7366.

[14] Erik Eotovos, Marcel Bodor, "DC–DC Resonant Converters for PV Applications", *IEEE Transactions on Power Electronics*, vol. 17, no. 3, 2013, pp. 5–15.

[15] Khandker tawfqie Ahmed, et al., "A Novel Two Switch Non-Inverting Buck-Boost Converter based Maximum Power Tracking System", *International Journal of Electrical and Computer Engineering*, vol. 3, no. 4, August 2013, [no pagination].

Chapter 4

Investigation of various solar MPPT techniques in solar panel

R. Divya and C.V. Pavithra

PSG Institute of Technology and Applied Research, Coimbatore,
Tamil Nadu, India

M. Sundaram

PSG college of Technology, Coimbatore, Tamil Nadu, India

CONTENTS

4.1 INTRODUCTION

Concerns about environmental consequences such as climatic change, global warming, fossil fuel depletion, and rising fuel prices are making renewable energy more appealing. Solar energy is one of the most promising renewable energy sources since it is abundant, pollution-free, and noise-free. PV generation systems, unfortunately, have two fundamental flaws: conversion efficiency in electric power generation is low, and the amount of electric power generated by solar modules varies constantly with weather conditions. Commercially accessible solar cells have an efficiency of roughly 38–43%, and solar systems have an efficiency of around 25–29% [1]. As a result, it is vital to employ effective approaches in order to maximize energy conversion efficiency while lowering costs.

The conversion of radiant (photon) energy from the sun to Direct Current (DC) electrical energy is known as photovoltaic (PV). Although PV power output remains low, ongoing efforts are being made to improve the PV converter and controller for optimum power extraction efficiency and lower cost. Maximum Power Point Tracking (MPPT) is a technique for tracking one maximum power point from an array's input and adjusting the voltage and current delivery ratio to get the most power possible. A maximum power point tracker, or MPPT, is basically an efficient DC to DC converter used to maximize the power output of a solar system. The first MPPT was invented by a Australian company called AERL way back in 1985, and this technology is now used in virtually all grid-connect solar inverters and all MPPT solar charge controllers. For maximal power extraction, a number of methods have been devised. Furthermore, the V-I characteristic of solar cells is non-linear and fluctuates with irradiance and temperature. In general, the maximum power point (MPP) is a single point on the V-I or V-P curve when the complete PV system performs at highest efficiency and delivers its maximum output power.

There are different algorithms in MPPT techniques. Constant voltage MPPT is one of the oldest algorithms which suffer many limitations. It just simply measures the ambient temperature of modules surroundings and adjusts the voltage accordingly. The advantage of this method is less requirement of electrical hardware for implementation and can provide acceptable power increases with little cost. However, solar irradiances impact on the MPP of a module is not observed in this method which leads to the lack of ability to truly find the maximum power point. Later on, methods like perturb and observe, hill climbing, and incremental inductance were introduced which is compared in the following chapters.

4.1.1 PV module

In order to raise the output current of a PV module, the cells are connected in parallel, however increasing the output voltage necessitates that the cells be connected in series [20]. In addition, a PV string is a system established by connecting PV modules in series, and a PV array is a system made by connecting PV strings in parallel [3–6]. Many methodologies have been developed in the literature to accurately examine the output characteristic of the PV system. These methods often emerge as analytical equations that are utilized to generate the equivalent circuit of a solar cell [9–19]. When analyzing

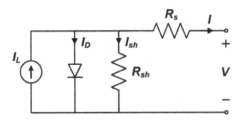

Figure 4.1 Analogous circuit model for a single-diode solar cell.

the performance of a PV system, it is preferable to utilize an equivalent circuit. Due to the balance of accuracy and simplicity, the single-diode model is the most recommended equivalent circuit model for modeling a solar cell [3, 20–22]. Figure 4.1 shows the analogous circuit model for a single-diode solar cell.

Equivalent circuit models define the entire I-V curve of a cell, module, or array as a continuous function for a given set of operating conditions. One basic equivalent circuit model in common use is the single-diode model, which is derived from physical principles [7, 8].

The governing equation for this equivalent circuit is formulated using Kirchoff's current law for current *I*:

$$I = I_L - I_D - I_{sh} \qquad (4.1)$$

Here, I_L represents the light-generated current in the cell,

I_D represents the voltage-dependent current lost to recombination, and

I_{sh} represents the current lost due to shunt resistances. In this single-diode model,

I_D is modeled using the Shockley equation for an ideal diode:

$$I_D = I_0 \left[\exp\left(\frac{V + IR_s}{nV_T} \right) - 1 \right] \qquad (4.2)$$

where *n* is the diode ideality factor (unitless, usually between 1 and 2 for a single junction cell), I_0 is the saturation current, and V_T is the thermal voltage given by:

$$V_T = \frac{kT_c}{q} \qquad (4.3)$$

where *k* is Boltzmann's constant $(1.381 * 10^{-23} \text{J/K})$ and *q* is the elementary charge $(1.602 \times 10^{-19} \text{ C})$

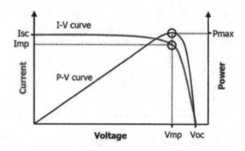

Figure 4.2 Typical I-V and P-V characteristics of a PV module.

Writing the shunt current as $I_{sh} = (V + IR_s)/R_{sh}$ and combining this and the above equations results in the complete governing equation for the single-diode model:

$$I_M = I_L - I_0 \left[\exp\left(\frac{V_M + I_M N_S R_S}{nN_S V_T} \right) - 1 \right] - \frac{V_M + I_M N_S}{N_S R_{sh}} \tag{4.4}$$

where I_M and V_M are the current and voltage, respectively of the module or array. Care should be taken when implementing model parameters, as they are either applicable to a cell, module, or array. Parameters for modules or arrays are strictly used with the single-diode equation for I, which is more commonly implemented form.

From the typical PV characteristics of a solar cell as shown in Figure 4.2, it is observed that at voltage 'V_{mp}' the power is at the maximum. This is the maximum power point of PV characteristics that we need to track using MPPT algorithm.

4.2 PROBLEM STATEMENT

The functioning principle of a MPPT solar charge controller is rather simple due to the varying amount of sunlight (irradiance) landing on a solar panel throughout the day, the panel voltage and current continuously changes. In order to generate the most power, an MPPT sweeps through the panel voltage to find the 'sweet spot' or the best combination of voltage and current to produce the maximum power.

The typical electrical characteristics of the module each consisting of 36 polycrystalline silicon solar cells, are given in Table 4.1. PV characteristics curves are obtained from the simulation for various temperature and solar irradiance. For the modeling of PV array, MATLAB Simulink is used. The typical V-I and P-V characteristics for various temperature and irradiance is shown in Figures 4.3–4.6. From the graph it is clear that the PV output power depends on two factors, namely solar irradiance and temperature. In

Table 4.1 Electrical characteristic of PV module

Parameter	Values
Maximum Power	60 W
Voltage at maximum power (V_{max})	17.1 V
Current at maximum power (I_m)	3.5 A
Short Circuit Current (I_{sc})	3.8 A
Open-Circuit Voltage (V_{oc})	21.1 V
Temperature Coefficient	(0.065±0.15)A/°C

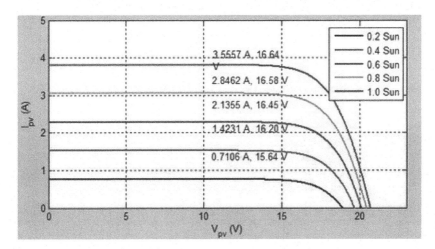

Figure 4.3 I-V characteristics at various solar irradiance.

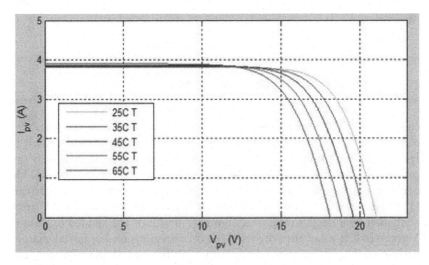

Figure 4.4 I-V characteristics at various temperature.

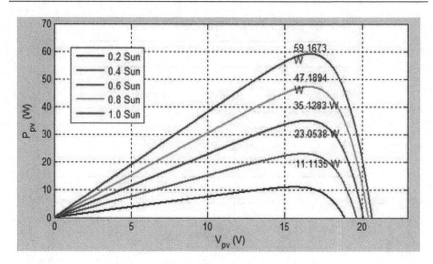

Figure 4.5 P-V characteristics at various solar irradiance.

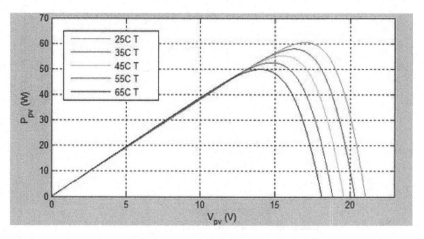

Figure 4.6 P-V characteristics at various temperature.

general, the output voltage and current is influenced by temperature and solar irradiance. That is higher the solar irradiance output current will be high and vice versa. From the I–V and P–V characteristics, there is only one optimum point which delivers the maximum power to the system. The energy produced by a PV array is not constant due to its non-linear nature [31]. In any event, partial shading occurs when some areas of the PV array are shaded by a neighboring tree, chimney, or cloud. The shaded portion of PV receives less solar intensity than the other regions under partial shading conditions. The electric power provided by the non-shaded PV modules would be absorbed by the shaded PV module in considerable amounts.

This is known as the hot spot problem, and it can harm PV cells. A bypass diode is usually linked in parallel with each PV module to provide an alternative channel under partial shadowing, hence preventing PV module damage.

4.3 METHODOLOGY

The operating point of PV varies from zero to the open-circuit voltage, from the I-V curve. The operating point does not always stay at maximum power due to variations in load. As a result, the PV does not provide the load with the maximum amount of available energy. To address this issue, the number of PV modules is increased, resulting in higher system costs and energy losses. To tackle this problem, an MPPT controller is used, which constantly searches for the maximum power point and makes the best use of the PV array. The MPPT continually tracks and adjusts the PV voltage to generate the most power, no matter what time of day or weather conditions. Using this clever technology, the operating efficiency increases and the energy generated can be up to 30% more compared to a PWM solar charge controller.

Several MPPT algorithms have been condition and partial shading condition [2, 23]. The conventional techniques failed to track the global peak whereas, Artificial Intelligence based algorithm have been developed to extract maximum power under any operating condition, including partial shading. The list of MPPT algorithm is shown as flowchart in Figure 4.7.

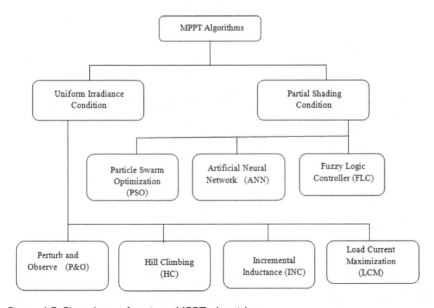

Figure 4.7 Flowchart of various MPPT algorithms.

In this chapter, the MPPT algorithm is organized into two categories. In the first part, the conventional MPPT algorithms which work satisfactorily under uniform irradiance are discussed. In the second part, the MPPT algorithms which work satisfactorily under both uniform irradiance and partial shading are explained.

4.3.1 MPPT algorithms for uniform irradiance

This section focuses on various MPPT algorithms under the uniform irradiance conditions. The Merits and demerits of each algorithm is discussed in detail.

4.3.1.1 Perturb and observe

The P&O algorithm is a relatively simple algorithm; as such it has a few drawbacks. The algorithm can be confused and track in the wrong direction, this can occur under fast changing irradiance conditions, the severity of this confusion depends on the P&O setup; i.e., step size and update frequency.

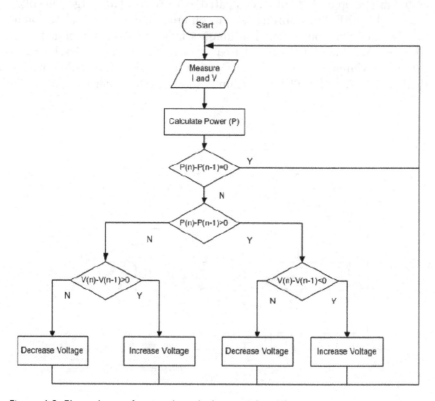

Figure 4.8 Flow chart of perturb and observe algorithm.

This algorithm perturbs the operating voltage to ensure maximum power. While there are several advanced and more optimized variants of this algorithm, a basic P&O MPPT algorithm is shown in Figure 4.8.

4.3.1.2 Hill climbing

Hill climbing, like the P&O algorithm, controls PV voltage to follow the ideal setting point (VMPP). The HC algorithm focuses on the duty-cycle perturbation of its power converter to find the MPP. The algorithm keeps track of the optimal point and updates it until the MPP, which is defined as $dp/dv = 0$, is discovered. $P(K)$, the current PV power, is constantly compared to P, the previous estimated value $(K - 1)$. The controller will measure the PV voltage and current again if the numbers are the same. If the current power is greater than the previous value, the slope is complimented. The switching duty-cycle of the power converter changes until the operating power oscillates at MPP [26, 27]. The drawback of this algorithm is that it fails to track the MPP under rapidly changing environmental conditions. The flowchart of the hill climbing algorithm is shown in Figure 4.9.

4.3.1.3 Incremental conductance

The flowchart of incremental conductance algorithm is shown in Figure 4.10. The equation for implementing the INC algorithm can be easily obtained from the basic power equation [24, 28]. The equation for power is given as

$$P = V \times I \tag{4.5}$$

Differentiating the above equation with respect to voltage yields,

$$\frac{dp}{dv} = I + V \times \left(\frac{di}{dv} \right) \tag{4.6}$$

The condition for the MPPT is that the slope $\frac{dp}{dv} = 0$.

$$\frac{dp}{dv} = 0 \, \text{at MPP} \tag{4.7}$$

$$\frac{dp}{dv} > 0 \, \text{left of MPP} \tag{4.8}$$

$$\frac{dp}{dv} < 0 \, \text{right of MPP} \tag{4.9}$$

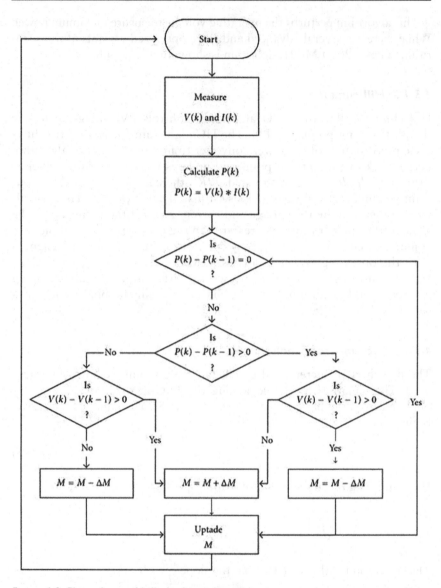

Figure 4.9 Flow chart of hill climbing algorithm.

In other words,

$$\frac{di}{dv} = -\frac{I}{V} \text{ at MPP} \tag{4.10}$$

$$\frac{di}{dv} > -\frac{I}{V} \text{ left of MPP} \tag{4.11}$$

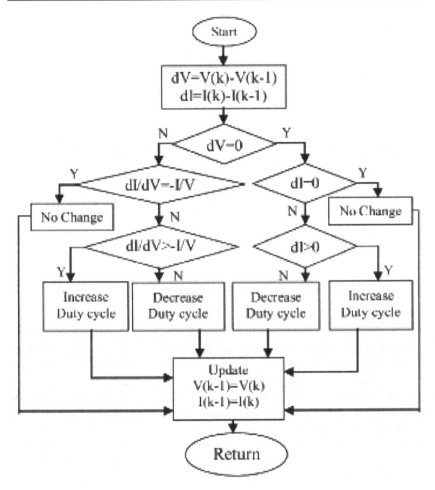

Figure 4.10 Flow chart of incremental conductance algorithm.

$$\frac{di}{dv} < -\frac{I}{V} \text{ right of MPP} \tag{4.12}$$

The disadvantage of IC is under partial condition, it is unable to track the global MPP. Hence, the control circuitry for IC is complex which results in high cost.

4.3.1.4 Load current maximization

In LCM we updated the fixed step strategy so that the step can be adjusted to increase tracking accuracy and convergence speed. Figure 4.11 depicts the flow chart of the LCM method with variable step size. In Figure 4.11,

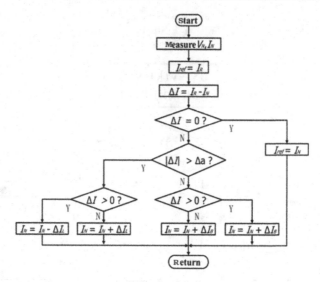

Figure 4.11 Flow chart of load current maximization algorithm.

the fixed current value at the photovoltaic cell's highest power point is I_R, the D-value between I_N and I_R is computed, and its absolute value is compared to the fixed current value Δa. When $|I_R - I_N| < \Delta a$, a small step ΔI_L is used as perturbation value. When $|I_R - I_N| > \Delta a$, a large step ΔI_B is used as perturbation value. When the environment of a WSN node changes often, the IR error value increases. As a result, the current value of the new maximum power point is utilized to cover the I_R when tracking each maximum power point. This method not only improves MPPT accuracy, but also convergence speed [30].

4.3.2 MPPT algorithms for partial shading condition

This section discusses in detail about the various algorithms for partial shading condition. In partial shading, there is possibility of having more than one maximum point in the photovoltaic output power curve. Hence, trained algorithms are required to derive the maximum power in this case. Particle Swarm Optimization (PSO) and Artificial Neural Network (ANN) are employed to extract he maximum power under partial shaded condition and the results are compared in this section.

4.3.2.1 Particle Swarm Optimization (PSO)

The flow chart for particle swarm optimization (PSO) algorithm is shown in Figure 4.12. PSO is a biologically inspired method based on animal social behavior. The process begins with a randomly distributed population of

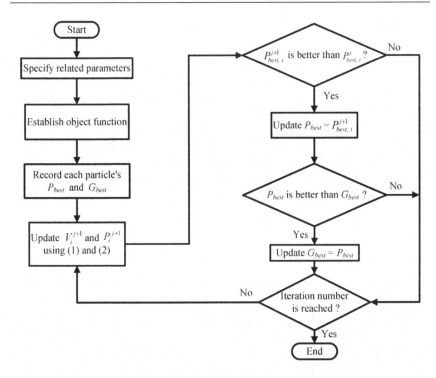

Figure 4.12 Flow chart of particle swarm optimization algorithm.

particles, each representing a potential solution. Every particle has its own velocity, which is changed by update equations that take into account the history of individual and collective experiences, i.e., each particle's experience as well as the experience of all particles in the population. The key goal is for the particles to evolve in such a way that they search the search space for the best answer. The algorithm uses a fitness function to evaluate each particle's performance during each iteration and modifies its velocity in the direction of its best performance until the instant (p_{best}) when each particle's best performance is reached (g_{best}).

Steps involved in PSO algorithm for extracting MPP are as follows:

1. Initially assume random particles in the search space, originated in PSO. The velocities of these particles are also randomly chosen.
2. Provide the solution of candidates to fitness function for evaluation and obtaining the fitness values among the particles.
3. Find out the particles global best and personal best among the entire particles.
4. Evaluate and update the velocities and position of each particle.
5. If convergence is achieved stop the search process if the condition is not satisfied raise the iteration count and once again start the evaluation of fitness process.

4.3.2.2 Artificial Neural Network (ANN)

The MPPT approach is also well-suited to artificial intelligence techniques like as neural networks and fuzzy logic [29, 33]. A computer model inspired by a biological neural network is known as an artificial neural network [25, 32]. In this approach, a neuron is a processing unit that linearly balances the inputs first, then calculates the sum using a non-linear function known as an Activation Function (AF), and lastly sends the results to the subsequent neurons as shown in Figure 4.13. Equation (4.13) gives the model of a typical neuron, where Z is the argument of AF.

$$Z = \sum_{m=1}^{M} W_m X_m + \alpha \tag{4.13}$$

Where, $X_1, X_2, \cdots, X_m$ are the m incoming signals, and $W_1, W_2, \cdots, W_m$ are the associated weights.

There are other training methods available, but the retro-propagation approach is the most well-known and widely employed. The method is trained by reducing the total error E, which is described by the following equation (4.14):

$$E = \frac{1}{2} \Sigma (O_n - t_n)^n \tag{4.14}$$

where O_n denotes the nth measurement read at the network's output and t_n denotes the nth target (the estimated output). As a result, each input/output pair is a sample. Using the equation [25], the back-propagation algorithm

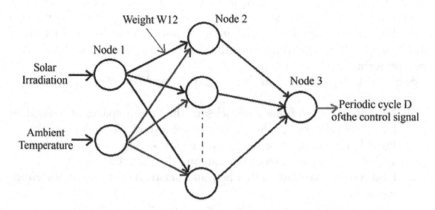

Figure 4.13 Multilayer feedback neural network.

calculates the error E and distributes the return of the output to the input neurons via hidden neurons:

$$\Delta \omega_n = \delta \Delta \omega_n - \eta \frac{dE}{d\omega} \qquad (4.15)$$

where ω is the weight between two neurons, $\Delta \omega_n$ and $\Delta \omega_{(n-1)}$ are the variations of these weights for n and $n-1$ iterations, δ is the duration of the regime and η is the training rate. The training rate determines the number of weight changes caused by the effect of total error.

The number of neurons selected in the hidden layer determines the degree of training. This number is calculated by the following empirical formula [25]:

$$N_h = \frac{1}{2}\left(N_1 + N_0\right) + \sqrt{N_E} \qquad (4.16)$$

N_h represents the number of neurons, N_1 represents the number of input neurons, N_0 represents the number of output neurons, and N_E represents the number of training samples. The training sample is continuously modified after each training by delivering all test data to the trained ANN model and recording the results to ensure network accuracy.

After that, the data is compared to the measurements. When the network converges, the performance factor is used to recreate the network's performance. Data validation is used to double-check the model's accuracy. If the network performs well in the sample and validation tests, we can assume that the network is capable of generating a fair periodic cycle.

4.4 SIMULATION AND PERFORMANCE COMPARISON

The simulation result for a 60 W PV array is presented in this section. The review has done a detailed simulation under uniform irradiance condition and partial shaded condition and the results are discussed in the following section. The simulation model in Figures 4.14–4.19 is based on Equations 4.1–4.4 whose typical I-V and P-V characteristics for various irradiances and temperature are shown in Figures 3.6.

4.4.1 Modeling and simulation of 60W PV array (Figures 4.14–4.19)

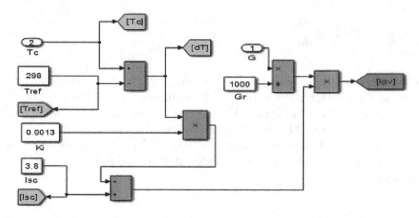

Figure 4.14 Simulation model for calculation of I_{pv}.

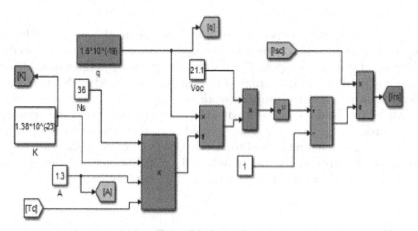

Figure 4.15 Simulation model for calculation of I_{rs}.

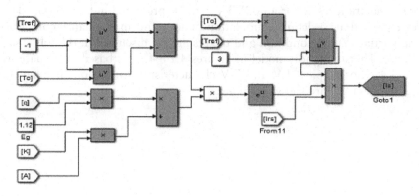

Figure 4.16 Simulation model for calculation of I_s.

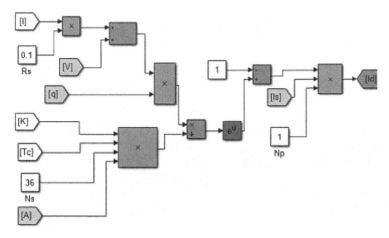

Figure 4.17 Simulation model for calculation of I_d.

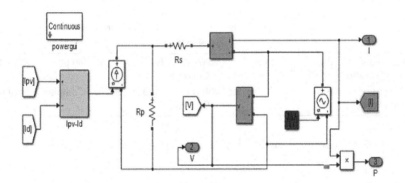

Figure 4.18 Simulation model for calculation of I.

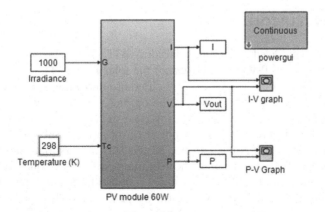

Figure 4.19 Simulink model of PV array subsystem.

Table 4.2 For uniform irradiance

MPPT algorithms	Criteria					
	PV array dependency	Sensitivity	Tracking parameter	Tracking speed	Complexity	Analog/Digital
Perturb and Observe	No	Moderate	Voltage, Current	Varies	Low	Both
Incremental Conductance	Yes	Moderate	Voltage, Current	Varies	Medium	Digital
Hill Climbing	No	Moderate	Voltage, Current	Varies	Low	Both
Load Current Maximization	No	Low	Current	Varies	Medium	Both

Table 4.3 For partial shading

MPPT algorithms	Criteria					
	PV array dependency	Sensitivity	Tracking parameter	Tracking speed	Complexity	Analog/ Digital
Particle Swarm Optimization	No	High	Voltage, Current	Fast	Simple	Digital
Artificial Intelligence	No	Moderate	Voltage, Current	Moderate/Low	Complex	Digital

4.4.2 Comparison of MPPT techniques for varying conditions of solar irradiation

This section gives the detailed comparative analysis of MPPT techniques under uniform irradiance and partial shading condition.

4.5 CONCLUSION

A PV string is modeled and simulated using MATLAB/Simulink in the proposed work. The I-V and P-V characteristics of the solar PV panel are determined at various levels of solar irradiances and different temperatures. It is found that higher the solar irradiance, higher is the output current and vice versa. Also, from the I–V and P–V characteristics, it is clear that, there is only one optimum point which delivers the maximum power to the system. But due to variation in load, the operating point does not always stay at maximum power point. Therefore, the PV does not supply the maximum power at all conditions. MPPT controller addresses the above mentioned problem by tracking the maximum power from the module under uniform irradiance as well as partial shading condition. In this chapter, various MPPT algorithms that work satisfactorily under both uniform irradiance and partial shading are discussed in detail.

REFERENCES

[1] S. Sengar, Maximum power point tracking algorithms for photovoltaic system: A review, *International Review of Applied Engineering Research*, 4(2), 147–154, (2014).

[2] T. Logeswaran, A. Senthil Kumar, A Review of maximum power point tracking algorithms for photovoltaic systems under uniform and non-uniform irradiances, *4th International Conference on Advances in Energy Research 2013, ICAER 2013, Energy Procedia*, 54, 228–235, (2014).

[3] N.A. Kamarzaman, C.W. Tan, A comprehensive review of maximum power point tracking algorithms for photovoltaic systems, *Renewable and Sustainable Energy Reviews*, 37, 585–598, (2014).

[4] N. Shah, C. Rajagopalan, Experimental evaluation of a partially shaded photovoltaic system with a fuzzy logic-based peak power tracking control strategy, *IET Renewable Power Generation*, 10(1), 98–107, (2016).

[5] V. Salas, E. Olìas, A. Barrado, A. Làzaro, Review of the maximum power point tracking algorithms for stand-alone photovoltaic systems, *Solar Energy Materials & Solar Cells*, 90, 1555–1578, (2006).

[6] P. Bhatnagar, R.K. Nema, Maximum power point tracking control techniques: State-of-the-art in photovoltaic applications, *Renewable and Sustainable Energy Reviews*, 23, 224–241,(2013).

[7] E.I. Ortiz-Rivera, F.Z. Peng, Analytical model for a photovoltaic module using the electrical characteristics provided by the manufacturer data sheet, in *IEEE 36th Conference on Power Electronics Specialists, 2087–2091*, (2005).

[8] A. Ortiz-Conde, D. Lugo-Muñoz, F.J. García-Sánchez, An explicit multiexponential model as an alternative to traditional solar cell models with series and shunt resistances, *IEEE Journal of Photovoltaics*, 2(3), 261–268, (2012).

[9] S.K. Hosseini, S. Taheri, M. Farzaneh, H. Taheri, An approach to precise modeling of photovoltaic modules under changing environmental conditions, *IEEE Electrical Power and Energy Conference (EPEC)*, (2016).

[10] W. Kim, W. Choi, A novel parameter extraction method for the one-diode solar cell model, *Solar Energy*, 84, 1008–1019, (2010).

[11] K. Ishaque, Z. Salam, H. Taheri, Simple, fast and accurate two diode model for photovoltaic modules, *Solar Energy Materials and Solar Cells*, 95, 586–594, (2011).

[12] S.K. Varshney, Z.A. Khan, M.A. Husain, A. Tariq, A comparative study and investigation of different diode models incorporating the partial shading effects, *2016 International Conference on Electrical, Electronics, and Optimization Techniques (ICEEOT)*, Chennai, 3145–3150, (2016).

[13] S. Vergura, A complete and simplified datasheet-based model of PV cells in variable environmental conditions for circuit simulation, *Energies*, 9(5), 2016.

[14] J. Cubas, S. Pindado, C. de Manuel, Explicit expressions for solar panel equivalent circuit parameters based on analytical formulation and the Lambert W-Function, *Energies*, 7, 4098–4115, (2014).

[15] M.G. Villalva, J.R. Gazoli, E.R. Filho, Comprehensive approach to modeling and simulation of photovoltaic arrays, IEEE.

[16] A.R. Reisi, M.H. Moradi, S. Jamasb, Classification and comparison of maximum power point tracking techniques for photovoltaic system: A review, *Renewable and Sustainable Energy Reviews*, 19, 433–443, (2013).

[17] K. Ishaque, Z. Salam, A review of maximum power point tracking techniques of PV system for uniform insolation and partial shading condition, *Renewable and Sustainable Energy Reviews*, 19, 475–488,(2013).

[18] F. Zhang, K. Thanapalan, A. Procter, S. Carr, J. Maddy, Adaptive hybrid maximum power point tracking method for a photovoltaic system, *IEEE Transactions on Energy Conversion*, 28(2), 353–360, (2013).

[19] N.S. D'Souza, L.A.C. Lopes, X.J. Liu, An intelligent maximum power point tracker using peak current control, In *2005 IEEE 36th Power Electronics Specialists Conference*, Recife, 2005, 172–177, (2005).

[20] N.S. D'Souza, L.A.C. Lopes, X.J. Liu, Comparative study of variable size perturbation and observation maximum power point trackers for PV systems, *Electric Power Systems Research*, 80, 296–305, (2010).

[21] S. Baraskar1, S.K. Jain, P.K. Padhy, Fuzzy logic assisted P&O based Improved MPPT for photovoltaic systems, In *International Conference on Emerging Trends in Electrical, Electronics and Sustainable Energy Systems (ICETEESES–16)*, 250–255, (2016).

[22] R. Abdul-Kalaam, S.M. Muyeen, A. Al-Durra, Review of maximum power point tracking techniques for photovoltaic system, *Global Journal of Control Engineering and Technology*, 2, 8–18 (2016).

[23] A. Chatterjee, A. Keyhani, Thevenin's equivalent of photovoltaic source models for MPPT and power grid studies, In *2011 IEEE Power and Energy Society General Meeting, San Diego, CA*, 1-7, (2011).

[24] Q. Mei, M. Shan, L. Liu, J.M. Guerrero, A novel improved variable step-size incremental-resistance MPPT method for PV systems, *IEEE Transactions on Industrial Electronics*, 58(6), 2427–2434, (2011).

[25] Z. Salam, J. Ahmed, B.S. Merugu, The application of soft computing methods for MPPT of PV system: A technological and status review, *Applied Energy*, 107, 135–148, (2013).

[26] Shah, N., Chudamani, R., A novel algorithm for global peak power point tracking in partially shaded grid-connected PV system, In *Proc. Int. Conf. IEEE Power and Energy Conf. (PECon)*, Kota Kinabalu, Sabah, Malaysia, 2–5 December 2012, 558–563, (2012).

[27] K. Ishaque, Z. Salam, G. Lauss, The performance of perturb and observe and incremental conductance maximum power point tracking method under dynamic weather conditions, *Applied Energy*, 119, 228–236, (2014).

[28] H.A. Sher, A.A. Rizvi, K.E. Addoweesh, K. Al-Haddad, A single-stage stand-alone photovoltaic energy system with high tracking efficiency, *IEEE Transactions on Sustainable Energy*, 8(2), 755–762, (2017).

[29] S. Hajighorbani, M. Radzi, M. Ab Kadir, S. Shafie, R. Khanaki, M. Maghami, Evaluation of fuzzy logic subsets effects on maximum power point tracking for photovoltaic system, *International Journal of Photoenergy*, 2014, (2014).

[30] N. Karami, N. Moubayed, R. Outbib, General review and classification of different MPPT Techniques, *Renewable and Sustainable Energy Reviews*, 68, 1–18, (2017).

[31] C.V. Pavithra, Vivekanandan Chenniyappan, Modified standalone single stage three port converter with domain distribution control for renewable energy applications, *Asian Research Consortium*, 7(3), 945–963, (2017).

[32] Suman Lata Tripathi, Mithilesh Kumar Dubey, Vinay Rishiwal, Sanjeev Kumar Padmanabhan, *Introduction to AI techniques for Renewable Energy System*. Taylor & Francis, 2021, ISBN 9780367610920.

[33] A. Inbamani, P. Umapathy, K. Chinnasamy, V. Veerasamy, S.V. Kumar, Artificial intelligence and Internet of things for renewable energy systems, *Artificial Intelligence and Internet of Things for Renewable Energy Systems*, 12, (2021).

Chapter 5

Real-time solar farm performance monitoring using IoT

E. Kannapiran, S. Jaganathan, and N.R. Govinthasamy
Dr. N.G.P. Institute of Technology, Coimbatore, Tamil Nadu, India

E. Kalaivani
Bannari Amman Institute of Technology, Erode, Tamil Nadu, India

CONTENTS

DOI: 10.1201/9781003302964-5

5.1 INTRODUCTION

The energy demand is constantly increasing, resulting in excessive consumption. The resources of fossil fuels is, in fact, critical to understanding the energy requirements of today's society. It is critical to develop more environmentally friendly, long-term solutions that are both efficient and beneficial to the environment. Solar energy is the most interesting renewable energy source because, due to significant cost reductions and technological advancements, it can bridge the gap between usage and production. This technology has evolved into a dependable source of information energy as a result of modern monitoring and control technologies. Smart grids use Information and Communication Technology (ICT) to increase sustainability, increase efficiency addition, and maintain a healthy balance between energy production and demand projections while consuming fewer resources.

Every human being nowadays requires electrical power to live a comfortable existence. Everyday power demands are on the rise, but alternative energy sources are declining daily. In order to meet the demand for power, multiple sources of electricity must be employed. There are two possible ways to produce electricity: non-renewable and renewable resources. Nature does not regenerate non-renewable resources, such as fossil fuels, coal, natural gas, and nuclear fuels after their initial use, but renewable sources, such as the sun, geothermal energy, wind energy, and tidal energy, may be used again, and always remain accessible. In this way, solar energy is viewed as an environmentally friendly source of energy.

Due to its abundant availability and low environmental impact, solar power has grown in popularity around the world. However, as conversion technology progresses, solar power generation becomes more affordable. Monitoring the solar system at the customer level is an urgent need at the moment. More appropriate energy sources may entirely replace on-renewable energy sources in the next years. Solar PV systems produce an adequate amount of power constantly. As a result, its performance must be observed instantaneously. The Internet of Things technology provides a solution for instantaneous monitoring of solar factors. An IoT device can interact with both the machine and the Cloud. We were able to wirelessly retrieve data from the cloud thanks to the deployment of IoT. In every solar power producing system, monitoring system parameters is critical. Real-time monitoring of numerous sensors detects crucial system constraints like current, voltage, irradiance, and temperature of the solar photovoltaic system. A monitoring system for photovoltaic panels isolated from the grid has several problems, such as tracking voltage, current, irradiance, and temperature. The solar system's real data must be synced regularly. For this node, MCU is utilized to communicate between devices and transport data to the cloud. Since the Solar PV form is connected to a cloud server via a LoRa IoT device, the suggested method can be adapted.

In a smart grid, renewable energy will be used. The IoT permits the internet to connect a huge amount of intelligently linked physical devices, permitting the latest technology of communication between things as well as people, data exchange for observing, and governing devices from everywhere in the world [3]. Furthermore, IoT applications enable machines or devices to communicate with one another without the need for human involvement [4]. The Internet of Things concept was created on the idea of connecting sensors and equipment from a single system to a larger network using wired or wireless nodes.

Wireless IoT solutions are popular because they eliminate the risks associated with wired systems. Each piece of equipment will be smart, autonomous, and remotely connected to meet future demands [3]. Biographers investigated methodological elements of IoT-enabled technologies, protocols, and applications.

The performance monitoring system bridged the gap between IoT and other evolving expertise like cloud computing, big data analytics, and fog computing. Photovoltaic monitoring systems are designed to deliver instantaneous information on a wide range of factors, containing energy potential, extracted energy, fault identification, past plant investigation, and related energy loss. Likewise, the information collected may be intended for preventive maintenance, early warning detection, and weather variance evaluation [2] goes into additional depth regarding the classification of photovoltaic power plant monitoring systems based on the internet expertise, information collection technologies employed, and monitoring system technique. Solar PV plants, charging stations for electrically powered vehicles [6], microgrids [7], and highway lighting, to name but a few applications, could benefit from monitoring tools. Two other important applications are water quality monitoring [8] and solar thermal station monitoring and management with solar collectors [9].

We investigated the wireless isolated monitoring connected to photovoltaic systems in the early 1990s, even though we are willing to take part in the photovoltaic component of solar energy. A thorough examination of solar plant monitoring systems was conducted [1] before moving on to discuss current and future system challenges and opportunities, as well as data collection systems' communication and storage. Several studies [10–13] have found that using cloud data recording and a monitoring system based on LabVIEW. PV performance indicators can be monitored and processed from anywhere. Aa low-cost Internet of Things-based solution was developed for a solar photovoltaic plant monitoring system that sends data via GPRS and a microcontroller [14].

It has been claimed that to monitor the parameters of solar household systems in real-time using an Arduino microcontroller and a 3G connection [15]. On the other hand, it has been proposed that an isolated monitoring system intended for rural solar photovoltaic systems based on GSM voice channels [16]. An experimental prototype of a solar PV plant monitoring system was created using the Internet of Things [17]. Furthermore, a low-cost IoT technique for remotely monitoring a solar system's Maximum Power Point (MPP) has been described [18]. A solar farm health monitoring system that monitors cord voltage, current, temperature, and humidity using eight photovoltaic panels and a validation technique have been developed [4].

Because solar panels are very sensitive to environmental characteristics such as irradiance, temperature, and humidity, meteorological data are considered crucial in evaluating the condition of a PV plant. As a result, to ensure proper operation, all PV systems must be monitored [19] by using CC3200 and ARM Cortex-M4 microcontroller. In the literature, there are numerous wireless resolutions for observing photovoltaic systems implemented with cheaper embedded boards like Raspberry Pi and Arduino. But the collective link with these explanations is the use of costly pyrometers to measure the irradiance of the photovoltaic system. The cheapest lux meter is smartly used in this project. The recommended monitoring system includes data collecting, handling, analysis, and display of PV plant performance. There is also an alert system in place, which is triggered based on the metrics that are measured. For the equipment and monitoring system, all of this is reasonably priced. The recommended solution, which makes use of the ESP32 board, is built around open-source and low-cost sensing, data collection, processing, and transmission technologies. Data-driven irradiance estimation was obtained using a pyrometer and lux meter to demonstrate the universal luminous value, as a result, the cost of the devices used was reduced by substituting a low-cost illuminance sensor for an expensive pyrometer.

The remainder of the chapter comprises: Section 5.2 deliberates the proposed IoT-based photovoltaic data monitoring system design; Section 5.3 describes the proposed system's hardware and software components; and Section 5.4 designates the result and discussion, and Section 5.5 investigates the Conclusion.

5.2 PROPOSED SYSTEM DESIGN

Because data in an IoT setup must be collected, processed, stored, and evaluated, this study creates the cheapest data collection for monitoring electrical and environmental parameters in a solar PV plant. The proposed PV plant monitoring system, the Node MCU unit, gathers and processes the information from different sensors before delivering the treated information to the cloud and servers through Wi-Fi.

The proposed block diagram of the IoT-based performance monitoring arrangement implemented in solar photovoltaic farm is illustrated in Figure 5.1. Transmission of treated information happens in dual stages: inter-integrated circuit protocol (I^2C) communicates between sensors and the controller, and the Wi-Fi protocol communicates between the controller and the cloud service application. The data collected by the various sensors

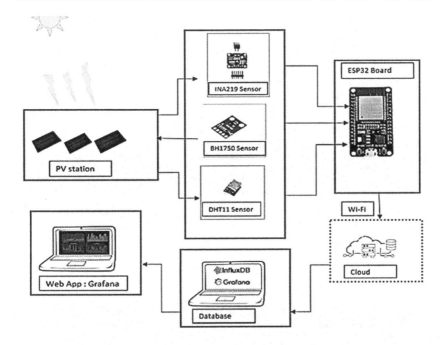

Figure 5.1 Proposed IoT-based solar form monitoring system.

are either kept locally or on the cloud (webserver). The server responds to an HTTP request sent by the client. The routing framework outlines how the various components of the web communicate with one another. A hardware arrangement used to collect the monitoring parameters from solar photovoltaic farm as wireless sensor node shown in Figure 5.2. A microcontroller collects the electrical and ecological parameters are required to evaluate the performance of the solar photovoltaic module, as described in Section 5.2, and sends them to the gateway via a LoRa wireless transceiver. Two more variables that can be measured are ambient temperature and orientation. The ambient temperature is included to provide more information about the backside temperature of the module and how these quantities vary depending on the environmental conditions. The orientation is included to help single and two-axis tracking systems understand irradiance readings.

5.3 HARDWARE DESIGN

The electrical and thermal characteristics of the solar panels used in the firm were discussed in the following sections. Solar panels are made of multicrystalline silicon. The layout of the solar farm in which the proposed monitoring system implemented as is shown in Figure 5.3.

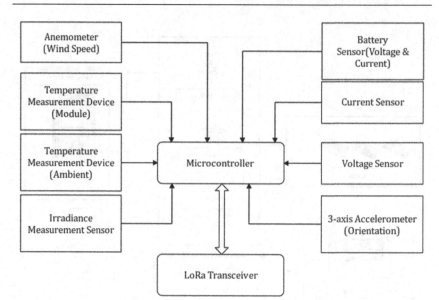

Figure 5.2 Block diagram of sensor arrangement of wireless data collection arrangement.

5.3.1 Electrical characteristics

✔ The proposed Solar Photovoltaic Module has a power rating of 300 Wp at STC.
✔ Solar Photovoltaic Module Type – Multicrystalline (HHV) & Tolerance: 0/+4.99
✔ Open Circuit Voltage (V_{oc}) of a Solar Photovoltaic Module in Volts: 45.5 V
✔ Short Circuit Current (I_{sc}) in Amps: 8.65 A
✔ Solar Photovoltaic Module Maximum Power Voltage (V_{mp}) in Volts: 36.5 V
✔ Solar Photovoltaic Module Maximum Power Current (I_{mp}) in Amps: 8.22 A
✔ Bypass Diodes – 3 Nos. Cells per Solar Photovoltaic Module: 72 Nos
✔ Solar Photovoltaic Module Efficiency: 15.43%
✔ Maximum system voltage is 1000 V direct current

5.3.2 Thermal characteristics

Influence of Variation in Temperature

✔ The minimum plant temperature under consideration is 5° Celsius
✔ The maximum plant temperature under consideration is 50° Celsius
✔ 300Wp has a maximum power voltage (V_{mp}) of 36.5 V

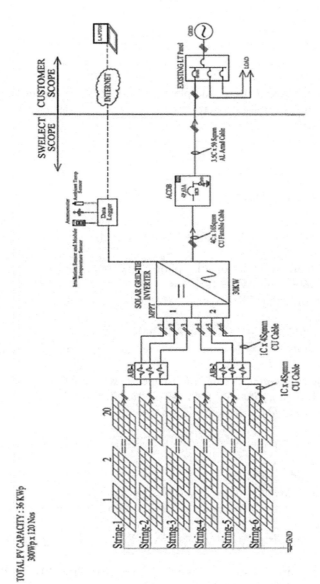

Figure 5.3 Layout structure of the solar form.

✔ Temperature coefficient is 0.45 percent of V_{mp} = 0.45% of 36.5 V = 0.16425 V

✔ The temperature rise in the plant will affects the operating voltage of the solar PV module selected is

= 50 – 25(STC) = 25°

= 25 × 0.164 = 4.1 V

= 36.5 – 4.1 V = 32.4 V

✔ The open circuit voltage (V_{oc}) of a 300Wp is 45.5 V

✔ Temperature Coefficient is 0.35% of V_{oc} = 0.35% of 45.5 V = 0.1547 V

✔ The temperature rise in the plant will affect the operating voltage of the solar PV module selected

= 25(STC) – 5 = 20°

= 20 × 0.1547 = 3.094 V

= 45.5 + 3.094 = 48.594 V

5.3.3 PV array design

✔ Solar plant capacity proposed for monitoring systems is 36 kWp

✔ Solar Photovoltaic Module designed to meet the specifications 300 Wp Capacity

✔ The required number of solar photovoltaic modules

= 36000/300 Wp = 120 Nos

✔ Count of Solar Photovoltaic Modules Connected in Series String (Minimum)

= 520 V / 32.4V = 16.02 ≅ 17 Nos

✔ Solar Photovoltaic Modules connected in series string (Minimum)

= 1000 V / 48.59V = 20.2 ≅ 20 Nos

✔ To reach anticipated rating and to operate within the Maximum Power Point Tracking range 20 modules are tied in series

✔ Total Solar Photovoltaic Modules Connected

= 6 × 20 = 120 Nos × 300 Wp = 36 kWp

✔ Maximum number of Power Point Tracking Modules per Inverter

= 2 Nos

✔ Total number of strings

= 120 Modules/20 Nos = 6 (3 strings per MPPT)

✔ There are three strings on each Maximum Power Point Tracking Module.

✔ Maximum String Current per Maximum Power Point Tracking Module

= 8 Amps × 3 string = 24 A (Inverter Capability: 30 A/MPPT)

A pictorial representation of the proposed performance monitoring hardware arrangement is illustrated in Figures 5.4 and 5.5. One ESP8266 is used for current sensing and one for voltage sensing in the proposed monitoring

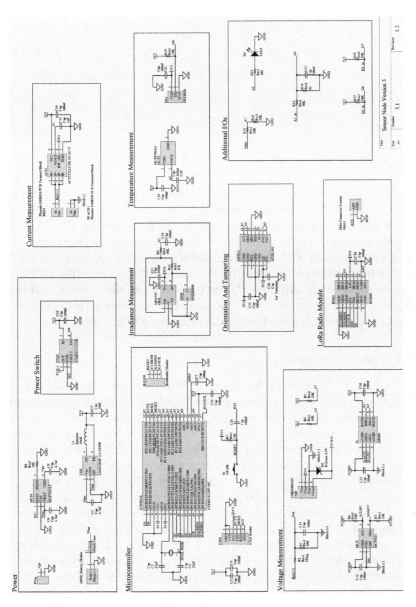

Figure 5.4 Schematic design layout of sensor node.

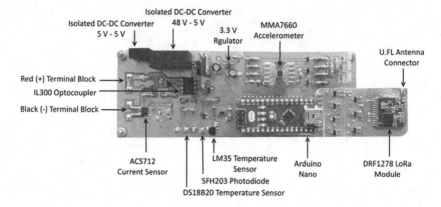

Figure 5.5 Sensor node assembly.

system. Temperature sensing is achieved by the DHT11 linked to the Node MCU ESP8266, which senses or receives the sun's radiation. Since ESP8266 requires a 5V DC supply, provided by an AC to DC converter.

An illustration of communication with the IoT is shown in Figure 5.6. After getting solar PV PCU data, the embedded gateway initiates a connection with the nos8266 node. Information is now collected from the PV PCU by the nos8266 node. Arduino IDE is used to write the C code, and the embedded hardware gateway receives the panel information from the solar PV system. It then starts sending the data serially over the app.

5.3.4 LoRa technology

The Low-Power Wide Area Network (LPWAN) technology is an adaptable, cost-effective, and low-power method for large, granular networks of wireless sensors. The LPWAN redefines how assets and processes can be monitored and managed remotely. Intended for IoT telemetry applications requiring periodic data transfers of modest volumes. LPWAN WSN technology includes LoRa (long range). It is a patented technology developed by Semtech, a company based in the United States. Chirp Spread Spectrum Technology is used (CSS). It also comes with Forwarding Error Correction (FEC). Long-range, secure transmission, and low power consumption are all advantages of LoRa. LoRa outperforms every existing technology in terms of power consumption, range, and cost. Lora WAN is a LoRa technology-based protocol developed by the LoRa Alliance. It is a free and open standard that covers the MAC, network, and application layers. We can achieve low power, wide area communication between remotely located sensor nodes and LoRa gateways for low latency and low bandwidth applications like data monitoring and control by combining LoRa with LoRa WAN protocol.

LoRa and LPWAN use unlicensed Industrial Scientific and Medical (ISM) bands to communicate. The operational frequencies for LoRa in India vary

Table 5.1 Comparison of performance characteristics of different wireless networks

WSN system	LTE	LPWAN	WI-FI	Zig Bee
Coverage	Huge	Huge	Less	Less
Range	High	High	Small	Small
Latency	Small	High	Small	Small
Bandwidth	200 kHz–900 MHz	500 kHz–900 MHz	2, 4, 3, 6, 5 & 60 GHz	<2.4 GHz
Power Consumption	Small	Small	High	Small
Topology	Star	Star & Mesh	Star	Point to point
Data Rate	Large	Small	Large	Small

from 865 MHz to 867 MHz. It is a license-free band designated by the Ministry of Communications and Information Technology of India for the use of low-power equipment. 863 MHz to 870 MHz is a license-free band in Europe. The operational frequency in this system was 866 MHz. The majority of LoRa networks use a star-of-star topology, which is made up of three different types of devices. End nodes, LoRa gateways, and LoRa Network servers are all included.

Many existing wireless technologies do not meet this requirement in large-scale industries and applications where a large area needs to be covered by WSN for efficient and long-range communication. Only LoRa technology is suitable for efficiently covering large areas with low-power sensor nodes (Table 5.1).

It is a good choice for large-scale deployments like this one because of its low power consumption, long range, low cost, and independence from existing infrastructure. When combined with a battery, these remote LoRa sensor nodes require very little power and support features like WOR (Wake on Receive) and Deep Sleep, making them even more power-efficient.

5.3.5 Communication pattern of LPWAN

Figure 5.6 Depicts the LoRa measurement message structure for IoT devices for solar PV system performance monitoring system. We can see here that communication may take place in two ways: first, data to the operator can be transmitted via a cloud platform or a mobile application on his or her smartphone. Following the completion of the different measures, a LoRa message is built and delivered to the gateway including the measurement data as well as additional data.

In monitoring systems with numerous sensor nodes, an acknowledge procedure is performed after each measurement message is broadcast to make sure that sensor nodes are not transferring measurement information at the same time described in the Figure 5.7. Following the delivery of a

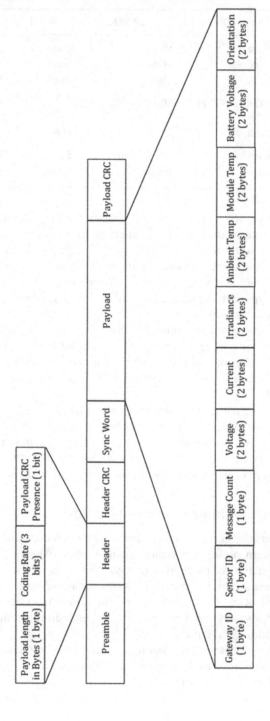

Figure 5.6 LoRa measurement message structure.

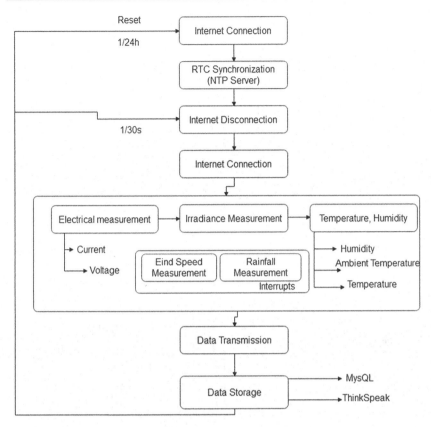

Figure 5.7 Process and communication sequence diagram of each iteration of routine time.

measurement message, a random interval period of 5 to 10 seconds is produced. After that, the LoRa unit is switched to nonstop receive mode. The segment is listening for any external messages in this mode. Upon receiving the LoRa communication comprising the Sensor ID and acknowledgment code, the sensor closes the load switch and enters sleep mode for the indicated measurement period.

The sensor node initiates new measurements, and the operation is restarted after the random interval period has expired if no communication with the Sensor ID and acknowledgment code has occurred. Sometimes the receiver did not receive the communication or acknowledgment message from the sensor node as shown in Figure 5.8. To avoid a temporal mismatch, the gateway time-stamp is used as the starting point for a new set of quantities. The time-stamp and real measurement time will be incorrect once the sensor node has sent the communication again after an interval. When two or more sensor nodes send measurements at the same time, a random interval of

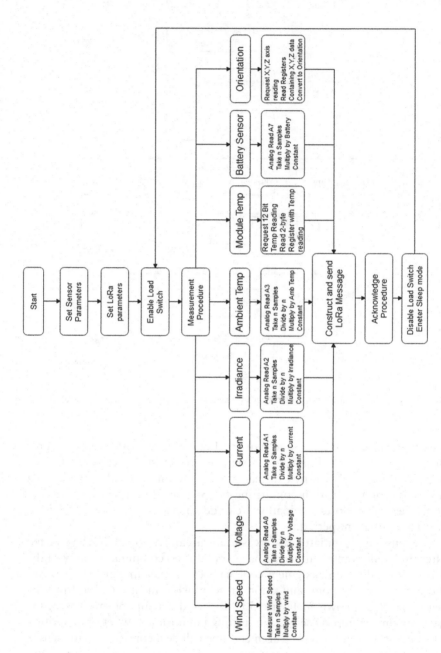

Figure 5.8 Sensor node operation flow chart.

5–10 seconds is set to ensure that each node sends the message again at a different time.

5.3.6 Gateway information management technique

A flow chart is shown in Figure 5.9 describes the gateway information management technique of software. The Gateway ID is set to 0xFF by default. The LoRa constraints are configured similarly to the sensor node settings. Communication between the LoRa modules and the base station must be successful in this case. A MySQL database is used, and it is hosted on the Google Cloud Platform.

When connecting to the MySQL instance, the IP address, a username and password are needed, as well as the database name. Each Sensor Node ID

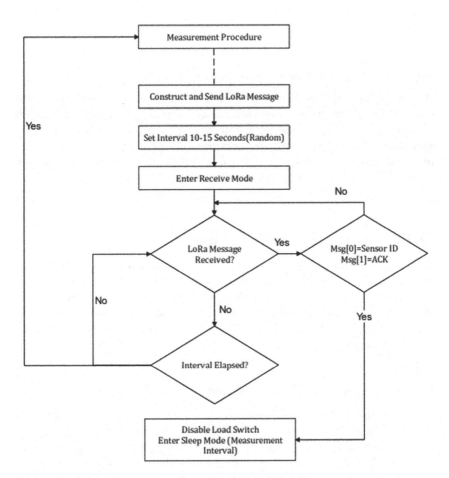

Figure 5.9 Acknowledge procedure of software flow.

has its table in the database. There is a column in each table for each specific measurement quantity.

The remaining information format is decoded and put in the proper variables after receiving a LoRa message with the Gateway ID in a first-byte position. The parameters are saved in the table associated with the Sensor ID, and a time and date stamp are appended. After the information is put in the relevant table, the acknowledge message sends by the gateway including the Sensor Node ID and the acknowledge code. This process will be repeated for receive mode elaborated in the Figure 5.10.

5.3.7 Functional requirements

To achieve the goals, the monitoring system's requirements were determined. Sensor nodes, gateway, an information storing capability (the cloud), in addition to a Graphical User Interface to make up the monitoring system.

5.3.7.1 Measurement of monitoring parameters

A solar PV module's performance is largely influenced by the electrical and environmental parameters captured by sensor nodes. Voltage, current, cell temperature, and irradiance are the factors. The sensor nodes are not necessary for measuring the closed-circuit current or open-circuit voltage, nor are they necessary for making the I-V curve. Taking these measurements may cause the PV module in an active solar system to behave differently than usual.

5.3.7.2 Collection of monitoring parameters

Sensor nodes must connect to a wireless network to provide measurement information to a central gateway. Wireless networks must have a low power consumption and a long range when used with solar PV power systems. Data from measurements must be saved in a database that the gateway can access remotely and quickly. A date and time-stamp must be included in each data packet.

5.3.7.3 Modeling measurement system

A model must be developed to analyze the data from the electrical and environmental measurements of each PV module. Electrical and environmental measurements must be incorporated into the model to predict and measure efficiency. In terms of user-defined thresholds, the model should be able to differentiate normal from problematic modules.

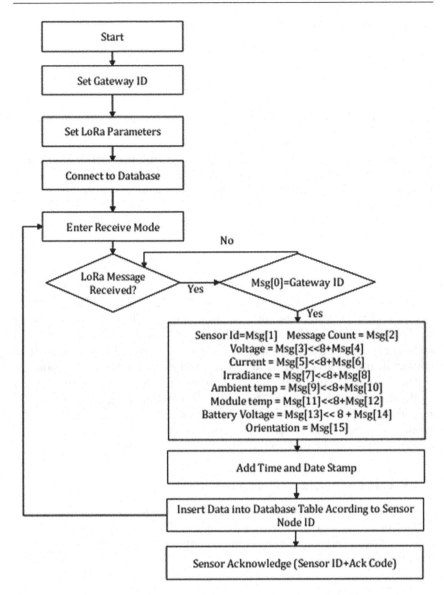

Figure 5.10 Flow chart for data handling procedure in a gateway.

5.3.7.4 Visualization of measured parameters

Visualizing measurement data (GUI) requires a graphical user interface. The layout of the PV system, as well as the positions of the numerous sensor nodes, must be displayed via the GUI. In addition to temperature, voltage, current, power, irradiance, and efficiency, the GUI should also display them. The health condition of each module must also be presented via the GUI.

5.3.7.5 Wireless monitoring and reporting

The graphical user interface (GUI) should be able to run on any computer connected to the Internet around the world. When PV modules or sensor nodes are somehow affected by the event, the GUI must notify the user.

5.3.8 Fault detection model

With the fault detection model, it is possible to determine whether a solar photovoltaic module isn't performing normally, and therefore, whether it is defective. When it comes to identifying defects in PV systems, there are three types of methods. Monitoring the temperature of the array is a key part of solar thermal monitoring. This involves tracking changes in the color of the modules as well as tracking hot spots. To use these techniques, thermal cameras and other modern monitoring devices are required.

In the recommended monitoring system, flaws can be uncovered purely by electrical methods since only electrical and environmental parameters have to be measured, as outlined in Section 5.4. Errors with solar photovoltaic modules will be broken down into three groups: fitting problems, fluctuations in electrical characteristics, and unintentional damage. The installation process may also cause physical damage such as cracks, along with alignment and inclination errors. There are several changes in electrical qualities, such as contact degradation, solder bond failures, and electrochemical corrosion.

There are many unintentional faults including shade, dust accumulation (soiling), abnormal aging, and mismatched manufacturing processes. A PV module fault detection model measures if there is a problem with the PV module and how much it underperforms because of the issue. In the fault detection process, the underlying cause of the fault cannot be identified.

5.3.9 Data analysis GUI application

The data analysis GUI program is developed in Python and makes use of frameworks and tools such as Matplotlib and Tkinter, among others. These are the two primary components of this interface – the Layout View and the Graph View.

a. Layout View

The layout view shown in Figure 5.11(a) depicts the locations of several sensor nodes as well as the overall structure of the solar PV installation. Each sensor node collects critical data, which is presented in the layout view. In addition to the ambient temperature, the module temperature, the string current, voltage, and power of the module, it also includes information about how efficiently the module operates. Taking a look at the layout view gives an idea of what the PV facility looks like today. Only the most current measurements are displayed in the layout view, which is refreshed every 10 seconds.

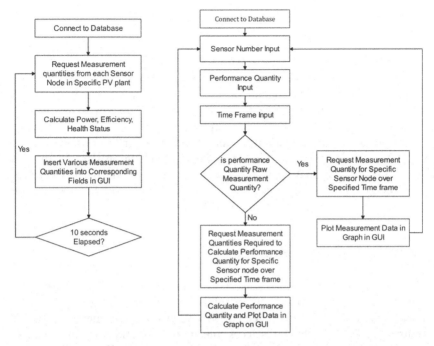

Figure 5.11 Flow diagram of (a) Layout view. (b) Graph view.

b. **Graph View**

A graph view shown in Figure 5.11(b) displays historical measurement data from each sensor node and users can examine any measurement quantity at any time. The user can also choose a time frame. The graph view aims to let the user observe and evaluate previous measurement data.

5.3.10 Software design

Programming with the ESP8266 is easy with Arduino, Ideal, or Esplora, as it uses a C-level language. Data transferred through the cloud can be accessed by an app. As a result, a valid login is necessary to access that data. As a consequence, a channel for reading data from the application is formed.

A. **Android App Studio**

Android App Studio, which was introduced on May 16, 2013, is an integrated development environment for producing Android mobile phone applications. This software program for Android phones is simple and quick to use. Apps may be written in Java, C, C++, and other programming languages.

B. **Think Speak Cloud**

Open-source IoT platform Think Speak uses HTTP Protocol to get data from sensors and collect or save it over the internet from devices connected to systems. It combines the data from sensors with items connected to systems using the cloud. It coordinates all sensor information and gives updates to the user through the Things Speak app. The user must first register with Things Speak Clouds before they can use it. Things Speak Clouds offers many channels for monitoring system characteristics from a remote device. An administrator or user can access and visualize the information in graphical form. Consequently, the user can access data from anywhere. Think of a channel management application where users may select whether or not they want to make their channel public or private by clicking on the add channel icon. Anyone may view their data in public mode by just entering their channel number. Reading and writing data from IoT device apps requires a key in the private model.

5.4 RESULTS AND DISCUSSION

This section describes the IoT-based real-time performance monitoring system implemented for the 36 kWp solar PV farm, as well as its hardware assembly and software arrangement to visualize the parameters shown in Figures 5.12–5.14. The data is accessible at any time and from any location, and it provides real-time status information on solar PV panels at regular intervals, resulting in a low-cost system. The Plots are generated on the mobile application interface by analyzing information collected from various sensors; the data is updated every one-minute interval, and the new data is saved on the IoT device.

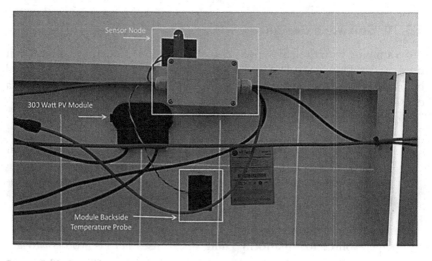

Figure 5.12 Bottom view for PV panel integrated with sensor node.

Figure 5.13 PV panel monitoring system with sensor arrangement.

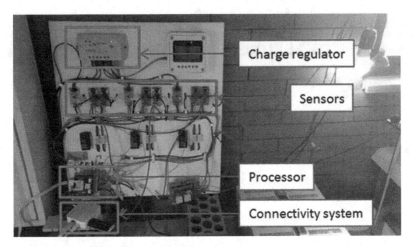

Figure 5.14 Completed product assembly of data monitoring system for PV solar plant.

A. **Voltage Plot**

As shown in Figure 5.15. The greatest value detected or created by the solar PV Panel is 16.02302 V, while the minimum value is 15.99 V, and the reading is taken by the voltage sensor between 12.00 pm and 12.52 pm.

B. **Current Plot**

As illustrated in Figure 5.16. This graph illustrates the value of current across the string, as measured by a current sensor between 12.00 and 12.52 p.m., with the biggest current being 1.57 amps and the smallest value being 1.48 amps.

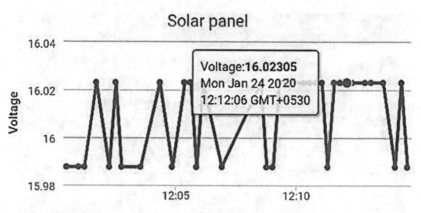

Figure 5.15 Voltage measurement by voltage transducer.

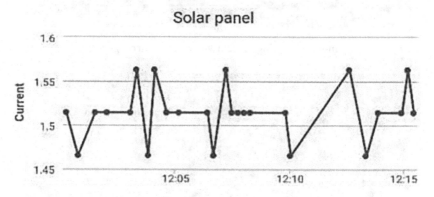

Figure 5.16 Current measurement by current transducer.

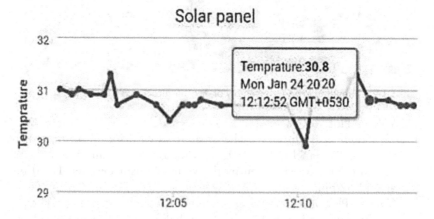

Figure 5.17 Temperature measurement by temperature sensor.

C. Temperature Plot

DHT11 senses the temperature on the surface of the PV panel, or the radiation received on it, by dividing the daytime by °C. As indicated in Figure 5.17. The maximum temperature achieved was 31.30°C, while the minimum temperature reached was 29.99°C.

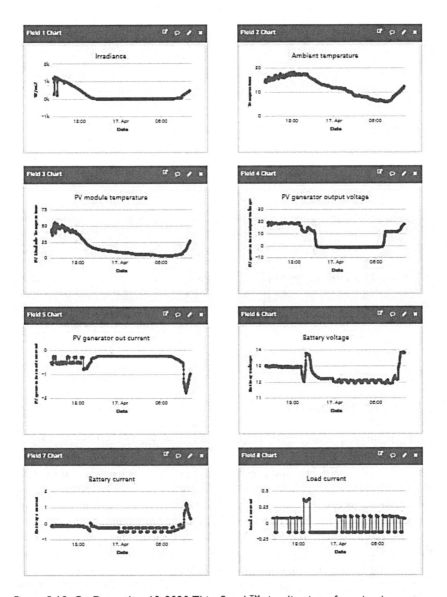

Figure 5.18 On December 13, 2020, ThingSpeak™ visualization of monitoring system.

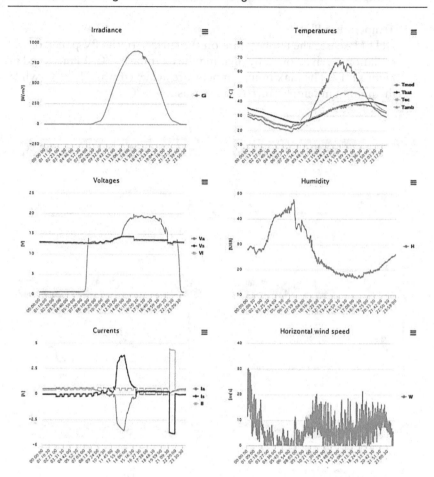

Figure 5.19 Browser version of performance data on 13/12/2020.

5.5 CONCLUSION

Esplora's platform was used to develop C programming language-based code for a node MCU so that a remote performance monitoring application for a Solar PV form could be performed by monitoring the parameters. With an Internet-of-Things-based MCU, this application has been successfully tested and proven to work. A solar panel monitoring application has been developed using Android App Studio to monitor voltage, current, temperature, and power. The node MCU sends data to the application over the internet and the application receives it. The IoT-based performance monitoring system is currently used to monitor a 36 kWp solar farm. It can be expanded to monitor many additional parameters in the future. This system is on-grid but can be adapted to an off-grid model without any modification.

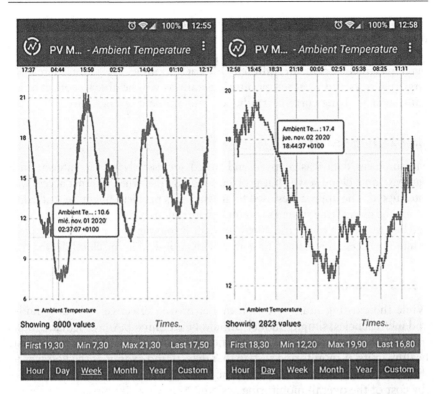

Figure 5.20 Screenshots are taken with the ThingView™ App on an Android Smartphone: On 1/11/2020 and 2/11/2020, the monitoring system recorded the weekly (left) and daily (right) ambient temperatures.

5.5.1 Recommendation

This section contains observations and suggestions for future work. Furthermore, problems that arose during the project are described, along with possible solutions.

5.5.1.1 Measurement

Several sensor nodes were designed to measure each variable associated with module performance, which included electrical and ecological variables. Thus, certain configurations result in the measurement of and sending of repetitive data for several quantities that have similar values. When a sensor is connected in a string, the same current feeds each one. As a result, measuring the current in a string using several sensors is no longer necessary. One primary sensor per string should monitor all electrical and environmental parameters; further sensors within the string should measure only voltage and module temperature.

The complete budget of the monitoring system will be significantly reduced with this configuration. Alternatively, the voltage and temperature of the module can be measured. Temperature and voltage are the two primary parameters that are different for each module. PV plants that have large inverters and weather monitoring stations will be able to collect other data, such as string current, irradiance, and ambient temperature.

5.5.1.2 Data collection

Measurement data was received and stored successfully by the gateway in a database on a cloud server, but problems would arise when internet access was interrupted. The approaches used to alleviate this problem might be studied to assure the monitoring system's continuing operation. It is advised that local storage be installed to retain measurement data if Internet access was interrupted. If Internet access is restored, the data will be published in the cloud database.

5.5.1.3 Modeling

While the existing fault detection model proved effective in finding faults and inefficiencies, simpler models should be examined. A problematic module's voltage was found to be substantially lower than that of adjacent, healthy units. A monitoring system that detects and examines the voltages of different solar photovoltaic within the solar farm can simplify and decrease the cost of the overall monitoring system.

5.5.1.4 Data visualization

To present plant data, a graphical user interface (GUI) application was based on the Python programming environment. This GUI was available only on computers running the Python environment. We might benefit from a web application accessed by any browser (PC, tablet, or smartphone).

5.5.1.5 Remote detection and reporting

The capability of detecting and reporting remotely would be greatly enhanced by a web-based application. It may also be considered to create a dedicated call center and/or response center capable of handling certain events and issues when they occur. Whenever a flaw appears, the relevant parties must be dispatched quickly and efficiently to correct it.

5.5.1.6 Future work

These project's sensor nodes will be deployed on a big scale in the future. The wide-scale deployment aims to monitor the temperatures of many PV modules in a utility-scale solar PV facility. The temperature data is primarily proposed to help in applying machine-learning techniques to develop a correct instant predicting model for a photovoltaic facility. A PV plant temperature map will be generated through this study, which will require detailed temperature data at the module level.

Other prospective uses for the monitoring system are being developed, such as a study subject examining various PV module technologies. In a range of test situations, the designed monitoring arrangement can deliver precise panel-level readings, allowing the comparison of various photovoltaic expertise and/or arrangements. Developing renewable energy sources will accelerate with the global shift toward "green" energy. In towns and countries, solar energy will increasingly become dominant, eventually dominating the electrical energy market.

Solar farms (PV plants) will play a more significant role in this scenario in the future. These companies must operate in the most cost-effective way possible. To accomplish this purpose, it will be required to assess the health of individual modules as well as bigger strings. In the absence of sufficient monitoring, the maintenance of modules, strings, and utility-scale plants is nearly impossible. Checking modules one at a time would take an unreasonable amount of time. This study indicates the importance of remote monitoring not only for the future of renewable energy but also for the present. This monitoring system would be able to be expanded to include other renewable energy sources without a problem.

Smaller hydropower facilities and wind farms might also be targeted for efficiency improvements. The work in this thesis might be the first, hesitant step toward an entirely new manner of sustaining solar energy facilities in the future. In the future, a standby arrangement established by using a micro-SD card would avoid information loss through the preservation of observed values while the system is disconnected. Furthermore, the connection mechanism was shown to fail under harsh situations (temperatures exceeding 40°C). Additionally, a more effective form of cooling system, as well as another 3G communication technology that can withstand greater temperatures, might be investigated. Future upgrades might include using solar cells to power the monitoring system. A data logger installed as a central node (by adding boards) could be used to monitor several solar array photovoltaic systems deployed close enough together due to the stackable hardware architecture. Technological developments in data communication networks are also anticipated.

REFERENCES

1 Zakaria, Z.; Kamarudin, S.K.; Abd Wahid, K.A.; Abu Hassan, S.H. The Progress of Fuel Cell for Malaysian Residential Consumption: Energy Status and Prospects to Introduction as a Renewable Power Generation System. *Renew. Sustain. Energy Rev.* 2021, *144*, 110984.

2 Reka, S.S.; Venugopal, P.; Alhelou, H.H.; Siano, P.; Golshan, M.E.H. Real-Time Demand Response Modeling for Residential Consumers in Smart Grid Considering Renewable Energy with Deep Learning Approach. *IEEE Access* 2021, *9*, 56551–56562.

3 Oprea, S.V.; Bara, A.; Andreescu, A.I. Two Novel Blockchain-Based Market Settlement Mechanisms Embedded into Smart Contracts for Securely Trading Renewable Energy. *IEEE Access* 2020, *8*, 212548–212556.

4 Narayan, S.; Doytch, N. An Investigation of Renewable and Non-renewable Energy Consumption and Economic Growth Nexus using Industrial and Residential Energy Consumption. *Energy Econ.* 2017, *68*, 160–176.

5 Afzal, M.; Huang, Q.; Amin, W.; Umer, K.; Raza, A.; Naeem, M. Blockchain-Enabled Distributed Demand Side Management in Community Energy System with Smart Homes. *IEEE Access* 2020, *8*, 37428–37439.

6 Han, J.; Choi, C.S.; Park, W.K.; Lee, I.; Kim, S.H. Smart Home Energy Management System Including Renewable Energy Based on ZigBee and PLC. *IEEE Trans. Consum. Electron.* 2014, *60*, 198–202.

7 Alzate, E.B.; Bueno-Lopez, M.; Xie, J.; Strunz, K. Distribution System State Estimation to Support Coordinated Voltage-Control Strategies by Using Smart Meters. *IEEE Trans. Power Syst.* 2019, *34*, 5198–5207.

8 Collotta, M.; Pau, G. A Novel Energy Management Approach for Smart Homes Using Bluetooth Low Energy. *IEEE J. Sel. Areas Commun.* 2015, *33*, 2988–2996.

9 Collotta, M.; Pau, G. An Innovative Approach for Forecasting of Energy Requirements to Improve a Smart Home Management System Based on BLE. *IEEE Trans. Green Commun. Netw.* 2017, *1*, 112–120.

10 Li, M.; Lin, H.J. Design and Implementation of Smart Home Control Systems Based on Wireless Sensor Networks and Power Line Communications. *IEEE Trans. Ind. Electron.* 2015, *62*, 4430–4442.

11 Bedi, G.; Venayagamoorthy, G.K.; Singh, R.; Brooks, R.R.; Wang, K.-C. Review of Internet of Things (IoT) in Electric Power and Energy Systems. *IEEE Internet Things J.* 2018, *5*, 847–870.

12 Shakerighadi, B.; Anvari-Moghaddam, A.; Vasquez, J.C.; Guerrero, J.M. Internet of Things for Modern Energy Systems: State-of-the-art, Challenges, and Open Issues. *Energies* 2018, *11*, 1252.

13 Tightiz, L.; Yang, H. A Comprehensive Review on IoT Protocols' Features in Smart Grid Communication. *Energies* 2020, *13*, 2762.

14 Lv, Z.; Qiao, L.; Verma, S.; Kavita. AI-enabled IoT-Edge Data Analytics for Connected Living. *ACM Trans. Internet Technol.* 2021, *21*, 1–20.

15 Polap, D. Analysis of Skin Marks through the Use of Intelligent Things. *IEEE Access* 2019, *7*, 149355–149363.

16 Kim, K.; Li, S.; Heydariaan, M.; Smaoui, N.; Gnawali, O.; Suh, W.; Suh, M.J.; Kim, J.I. Feasibility of LoRa for Smart Home Indoor Localization. *Appl. Sci.* 2021, *11*, 415.

17 Zainab, A.; Refaat, S.S.; Bouhali, O. Ensemble-based Spam Detection in Smart Home IOT Devices Time Series Data Using Machine Learning Techniques. *Information* 2020, *11*, 344.

18 Makhanya, S.P.; Dogo, E.M.; Nwulu, N.I.; Damisa, U. A Smart Switch Control System Using ESP8266 Wi-Fi Module Integrated with an Android Application.

In *Proceedings of the 2019 IEEE 7th International Conference on Smart Energy Grid Engineering (SEGE)*, Oshawa, ON, 12–14 August 2019; pp. 125–128.

19 Sarah, A.; Ghozali, T.; Giano, G.; Mulyadi, M.; Octaviani, S.; Hikmaturokhman, A. Learning IoT: Basic Experiments of Home Automation using ESP8266, Arduino and XBee. In *Proceedings of the 2020 IEEE International Conference on Smart Internet of Things (SmartIoT)*, Beijing, China, 14–16 August 2020; pp. 290–294.

Chapter 6

Solar energy forecasting architecture using deep learning models

R.R. Rubia Gandhi, C. Kathirvel, and R. Mohan Kumar
Sri Ramakrishna Engineering College, Coimbatore, Tamil Nadu, India

M. Siva Ramkumar
Karpagam Academy of Higher Education, Coimbatore, Tamil Nadu, India

CONTENTS

6.1 INTRODUCTION

Solar based power generation can be easily adopted in most of the places due to its low installation and maintenance cost. This power generation using panels is not always stable, which affects forecasting of solar power generation data. The stability factor is affected by different atmospheric conditions. An unstable power generation in any of the circumstances causes the grid instability. This serious issue can be addressed using machine learning

DOI: 10.1201/9781003302964-6

techniques for forecasting the solar power output. IoT combined with the AI techniques adding a helping hand in solving this kind of issue in most of the cases. The rapid development in IoT and AI techniques so far paved a way for the improving the accuracy in forecasting, power generation, energy consumption, and anomaly detection.

The relevant areas which have been reached popularity using this AI based forecasting techniques are cyber-physical systems, air quality forecasting, digital twinning etc. This type of forecasting is carried out in two phases.

- Clustering: the original dataset will be divided into different group of datasets using K-means clustering algorithm
- Forecasting: machine learning techniques like ANN, SVM, decision trees are performed to do the forecasting

The most used techniques like Convolution Neural Network (CNN), Long Short-Term Memory (LSTM) and Deep Belief Networks (DBN) are mostly used for the solar power generation forecasting.

Table 6.1 Algorithms commonly used for prediction

Researcher	Proposed method	Methodology
Yang et al.	Discussed a solar power forecasting based on deep learning weather types for hybrid system with a time step at each hour	Using self-organizing map learning vector quantization fuzzy method for classification, training and prediction
Han et al.	Discussed multi-model alternative internal solar power prediction	Seasonal behavior of the solar power Accurate deviation in PV power prediction
Denver et al.	Proposed a SVM based GA	Taken a weather dataset using an SVM classifier Optimized by GA using ensemble technique
Malvoni et al.	Proposed to use principal component analysis, wavelet decomposition	GLSSVM used for the time series forecasting method LS-SVM and GMDH used to measure the weather data
Webin et al.	Proposed a data mining approach	K-means algorithm used for clustering two categories ANN is used for wind forecasting
Aka et al.	Proposed ML approach	A short span of wind speed time series prediction Extreme learning machines and multi-objective genetic algorithms combined with the nearest neighbors approach for estimating prediction intervals of wind speed
Ren et al.	Proposed a SVM based GA	Empirical model decomposition using SVM

Table 6.2 Various deep learning models for energy forecasting

Author	DL models	Purpose
Rodrigo et al.	Used deep learning approach Evaluated the system with three algorithms fully connected, CNN, LSTM	For the analysis of monthly power consumption of accompany in Brazil
Amarasinghe et al.	Used CNN Validated the system with benchmark dataset of power consumption with a normal residential consumer	To forecast the energy load data values for single building
Kuo et al.	Used deep Convolution Neural Network Comprises of six layers: three convolution and three pooling	Prediction of coastal area data for solar
Yubo et al.	Used deep belief network The model consists of multilayer perception and SVM	Prediction of Solar and wind power
Huai et al.	Used deep learning approach Model is based on wavelet transformation and CNN	Prediction of solar power forecasting
Wan et al.	Used deep feature learning A DBN approach for regression The researcher compares the model with a hidden layer neural network	Wind speed forecasting

6.2 DEEP LEARNING APPROACH

Deep Learning is applied for various areas like analyzing imaging, speech recognition, and text mining. This is given by multilayer representations and hierarchical architectural features. Most commonly DL is applied for classifications and regression models [17]. The different models like RNN, CNN, deep belief network, and stacked autoencoder are presented in various research.

- RNN is always applied for time based serial data problems. This method tries to learn from the previous data inputs available in the memory
- CNN is composed of three layers namely, convolution layer, pooling layer, and fully connected layer
- Deep belief network is stacked up various hidden layers. It is a layer-by-layer connection not with the units inside the layers. It is composed of unsupervised networks similar to general stochastic neural network which tries to learn the model with some set of inputs
- Stacked autoencoder is composed of various autoencoders. It consists of two stages namely encoding and decoding stage

6.3 A SAMPLE MULTILAYER ARCHITECTURE

A new approach has been proposed by an author. The architecture comprised of multilayer such as IoT layer, DL model layer, and big data layer, to enhance the integration of renewable energy systems. A data miming process has been carried out for the renewable energy block.

- Renewable energy layer provides information about wind, solar, and performance parameters data. Simply put, this layer includes all the responsibilities about the renewable sector. Any problems that occur with the renewable need to be addressed in this layer
- In IoT layer, the information regarding the program configurations, IoT devices, and other circuit driver's communications are addressed in this layer
- In big data layer, the processes like the data collection, data processing, and data storing all these are involved and maintained here
- In deep learning layer, the major work of prediction occurs. It analyze the data coming from renewable layer, IoT layer

The IoT layer DL techniques which has wide focus in identifying the stages like approximation in implementing the stages, accuracy in predicting the power generation, giving a less convergence time and considering the uncertainty conditions in the forecast [18].

In this architecture, the IoT layer takes the data from the renewable layer. The data captured includes the power generating devices and the various factors affecting the power generating wind, solar, weather, and

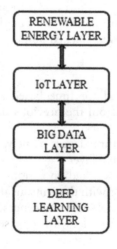

Figure 6.1 Multilayer architecture.

temperature etc. In line, the big data layer stores the captured data obtained from the IoT layer in $Data_{1.....n}$. After this each data is assigned to a node like $data_i$ to $node_i$. The author has implemented a HADOOP system for storing the data [18]. The file is of gigabyte or terabyte size. The adopted architecture takes large scale data files which capture the data geographically from the environment. DL layer consists of deep learning algorithms like algorithm 1, 2, 3. Each DL model uses the data from different set data nodes available in the big data layer. The created model is used for prediction.

6.4 DATA MINING PROCESS FOR RENEWABLE ENERGY

Data mining is done by the data analyst community. It involves various steps like data collection, data exploration, modeling the data and deployment of models.

The cross industry standard process for data mining is known as CRISP-DM which is a standard process used for data mining. An extended version of CRISP-DM is considered for renewable sector data mining here. This makes the process simple to digitalize the solar power forecast dataset. The extended data mining process includes:

- Determining and understanding the objective of the system needs to be modeled
- Data acquisition in which the IoT layer captures the data to create an input dataset
- Modeling is done by using any one of the best DL algorithm
- Evaluation is done to validate the usability
- Deployment phase is the final phase where the end user can use the proposed system

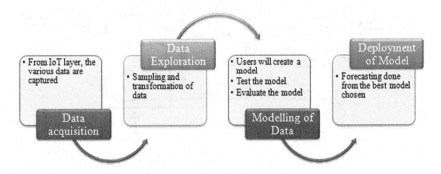

Figure 6.2 Data mining process.

6.5 SOLAR FORECASTING ARCHITECTURE

Reviewing the framework is divided into three phases. In the first phase, trained data is divided into four clusters with usage of SOM algorithm which includes the data of four seasons. In the second phase, model is built using AM, CNN, and LSTM. The third phase is forecasting in which appropriate prediction model need to be selected and processed.

Looking at a series of research relevant to this domain, it was found that a combination of CNN and LSTM is producing better results in solar forecasting. The process steps involve:

- The data collected is processed using clustering techniques to feature out accurate results
- CNN and LSTM combined DL framework is adopted
- For further enhancement, finally an AM technique is adopted

To predict the best technique, much research proposed testing results. The prediction method proceeds with four main streams:

1. Data Collection
2. Data preprocessing
3. Training the model and parameterization
4. Performance metrics

Finally, model testing and validation is carried out.

The sample architecture is provided in Figure 6.3 for solar forecasting.

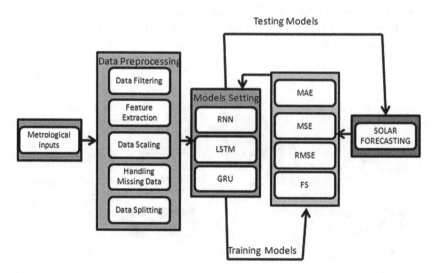

Figure 6.3 Method for solar power prediction.

6.5.1 Data collection

The solar power generation seems to be useful paradigm to achieve a clean and non-polluted power generation which is having a potential of unlimited high-power extraction. The mail difficulty in achieving this is availability of clear sunny day for solar power forecasting. It must be solved by identifying with proper technique for forecasting. By choosing one of the innovative approaches known as Myceilsi-Markov with the experimental verification, the discussion proceeds here.

With the help of extensive computations, the researchers have compiled a reconciliation to improve the forecast significantly. Using regression model, the optimal prediction of solar forecasting was done using various important factors that are highly dependent for the process [4]. Overall results infer that the proposed approach has given a satisfactory result compared with conventional techniques. A researcher has implemented the Gaussian regression technique along with CNN produces accurate results through a large experimental validation [8].

6.5.2 Dataset

Data was collected by UMASS University. Using sensor traces, the data set has been compiled, also consisting of weather data. The data has been collected from Amherst, Massachusetts weather station [1]. It was collected every 5 minutes, and parameters such as UV level, rainfall, humidity, dew, temperature, and atmospheric pressure. This dataset can be downloaded from the UMASS website and this dataset is loaded to the python version for feature extraction and then loaded into the corresponding repository.

Once dataset is collected, processing need to be done for the validation. This comprises of five processes.

--Timestamp---	Temp	Chil	HIndex	Humid	Dewpt	Wind	HiWind	WindDir	Rain	Barom	Solar	ET	UV
-------------	----	-----	-------	------	----	------	------	--------	-----	------	------	-----	----
20060202 16:55	44.9	44.9	44.9	60	31.9	3	3	180	0.00	30.071	-100000	0.000	-
100000.020060202 17:00	44.9	44.9	44.9	61	32.3	1	3	158	0.00	30.075	-100000	0.000
-100000.020060202 17:05	44.8	44.8	44.8	61	32.2	1	3	158	0.00	30.075	-100000	0.000
-100000.020060202 17:10	44.8	44.8	44.8	62	32.6	1	2	158	0.00	30.075	-100000	0.000
-100000.020060202 17:15	44.8	44.8	44.8	61	32.2	2	5	135	0.00	30.074	-100000	0.000
-100000.020060202 17:20	44.8	44.8	44.8	61	32.2	3	6	135	0.00	30.074	-100000	0.000
-100000.020060202 17:25	44.8	44.8	44.8	61	32.2	3	6	135	0.00	30.074	-100000	0.000
-100000.020060202 17:30	44.8	42.1	44.8	61	32.2	5	10	135	0.00	30.066	-100000	0.000
-100000.020060202 17:35	44.6	41.2	44.6	61	32.0	6	9	135	0.00	30.066	-100000	0.000
-100000.020060202 17:40	44.6	42.6	44.6	61	32.0	4	7	135	0.00	30.066	-100000	0.000
-100000.020060202 17:45	44.5	42.5	44.5	61	31.9	4	6	135	0.00	30.069	-100000	0.000
-100000.020060202 17:50	44.5	44.5	44.5	62	32.3	3	7	112	0.00	30.069	-100000	0.000

Figure 6.4 UMASS dataset.

6.5.3 Data preprocessing and feature selection

Data preprocessing and feature selection seems to be the most important part in the deep learning approaches. For making the data collected into a useful formatted data, data processing must be done. While collecting a data for a specific purpose, other irrelevant data not useful for performing the task must be removed. This can be sorted out by adopting a proper classification algorithm.

Then the feature selection which provides the accurate features that will add the advantage for the learning models. This will enhance the accuracy and reduces the costly computation. Procedures are needed in preprocessing the data and feature selection (see Figure 6.5).

The data source taken for the task may consists of unwanted features that may misguided the model to not to train accurately in predicting the forecast. Here, four different features are in play: wind pressure, temperature, humidity, and solar energy. These features are correlated for predicting the output. It is important to select an accurate model for accurate prediction. The features may present in different scales. To ensure all the featured data to be present in the same scale, MIN –MAX transformations of 0's and 1's is done. This will remove the unwanted noise signals and other disturbances and makes the model to be simplified learner for the task execution. Another important feature is all the time-series domain need to be predicted the present data which has been in the dataset. Most importantly, the model should try to predict the future data for an efficient and optimistic solution.

In some of the training models, data preprocessing, data processing, and data postprocessing has been carried out.

Data Cleaning: the prior step in data preprocessing is cleaning the data of the anomalies from the respective dataset.

Normalization: the next step is preprocessing. Here, all the data are scaled between 0 and 1. These data will bring into line the entire probability of the distributed input terms.

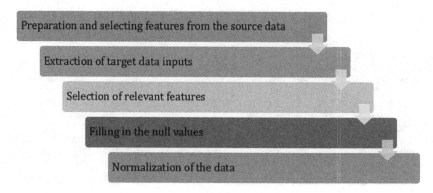

Figure 6.5 Procedures in data processing using ANN model.

Normalization equation is given by

$$y = \frac{\left(y_{max} - y_{min}\right)\left(x - x_{min}\right)}{\left(x_{max} - x_{min}\right)} + y_{min} \qquad (6.1)$$

In the above equation, y represents the normalized value of the input, y_{max} is 1, and y_{min} is 0; x has real values.

6.5.4 Model training and parameterization

When a model is taken for the training, each model needs to be parameterized during its phase of training for predicting accurate results. After the parameterization is done, a optimizer need to be adopted for further better training [22]. ADAM optimizer seems to be better than the stochastic optimizer used in other empirical models. This causes the model to learn quickly. Then the tanh activation function is chosen for the good fitting of the model training.

ANN model is chosen for training the data. After normalization, the data will be split into three datasets:

1. Training
2. Testing
3. Validation

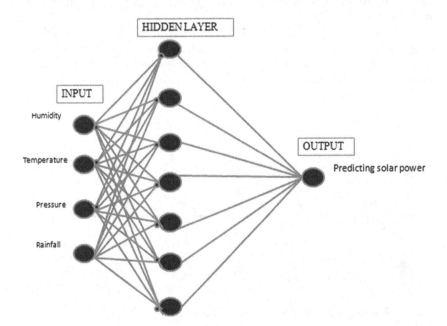

Figure 6.6 ANN training model.

Once this process is done, the kernel LSTM model is used to predict the solar power forecasting technique most efficient for reducing the root mean square error and main bias error.

6.6 SOLAR POWER FORECASTING PREDICTION TECHNIQUES

In reviewing the articles related to solar power forecasting, three different models has been frequently implemented to most of the data set and analyzed. They are RNN, GRU, and LSTM.

6.6.1 RNN

Recurrent Neural Network has been adopted for areas such as image processing, speech recognition, and uncertainty prediction. Whenever a situation of uncertainty or random behavior approaches the RNN based system, it is able to send the information to its internal memory which capable to predict the random sequence of the data approaching [21]. The internal memory stores the information about the previous calculated value. The below figure shown the architecture of RNN where feedback from other neurons is given to the hidden coming from the previous period multiplied by weight W. When the network is unfolded horizontally, from that it is found that the hidden neuron takes the input from the previous step.

When Comparing the RNN, a DRNN network is better in prediction of deep transition. But this deep transition increases the complexity in training. This can be overcome by having shortcut connections providing shorter paths, and bypasses the middle layer [3]. Overall, the gradient of RNN in

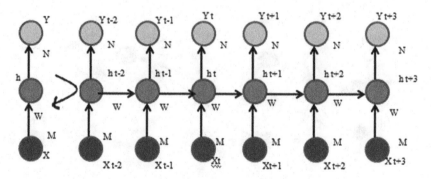

Figure 6.7 RNN architecture unfolded toward right.

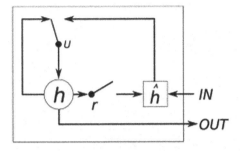

Figure 6.8 GRU cell structure.

tracking the long-term memorization is very difficult; this can be provided by LSTM technique [16].

6.6.2 GRU

The most recently adopted deep learning techniques in predicting solar power forecasting are gated RNN, as it is discussed above it has the type of model specially designed which gives promising results in predicting the sequential data with its ability to bypass the long-term data from its previous computation as from the inputs. Cho et al. have proposed this GRU technique. This network has multiple cells that selectively memorize or transfer. The architecture is simple compared to LSTM.

GRU means Gated Recurrent Units. It has two gates in the architecture for transferring the information. The most important feature of gated RNN is it will accurately understand the system and predicts the output sequentially with respect to time.

These GRU cells are similar to the LSTM technique which also has two gates. But the main difference is GRU cells contain only two gates. LSTM contain three gates. GRU requires more computations. It has reset (r), update (u), h represents present output, and h coefficients represents previous outputs as shown in Figure 6.2. The update gate will have the storage of information need to be remember. Reset gate will acknowledge whether any new information can be added to the previous state.

6.6.3 LSTM

Reviewing many research articles it is found that the deep learning-based training model can be proposed for execution of tasks in distributed and parallel mode. A distributed and parallel model for LSTM technique is used for solar power forecasting. As discussed previously in GRU architecture, LSTM model has three gated in the architecture. It is also a type of recurrent

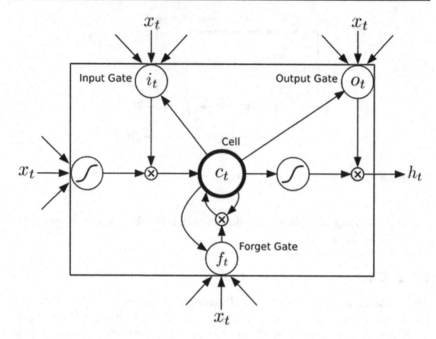

Figure 6.9 Structure of LSTM.

neural networks. Same kernel steps have been included for using the LSTM technique [4]. As discussed in RNN, long time prediction cannot be accurate and, in some cases, it is not possible. While adopting the LSTM technique, it is proposed to learn long-term dependencies and able to solve the vanishing and exploding problems which will result in RNN architecture.

It is composed of three gates. It is given by forget gate, input gate, and the output gate. The memory cell present in the structure records the historical information from the three gates in the present time. The gate acquires the values from the range [0, 1]. The gate activation function is given by

$$\text{forgetgate} = \sigma(W_f.[h_{t-1}, x_t] + b_f \tag{6.2}$$

$$\text{inputgate} = \sigma(W_i.[h_{t-1}, x_t] + b_i \tag{6.3}$$

$$\text{outputgate} = \sigma(W_o.[h_{t-1}, x_t] + b_o \tag{6.4}$$

Sigmoid Function: σ is a non-linear activation function which maintains the values between 0 and 1. This helps the model to forget or update the data. If the product results in 0, it forgets the data, if it results in 1, the model will learn and remember the information.

Tan h Function: Tanh is also a non-linear activation function. It maintains the values from the range −1 to 1.

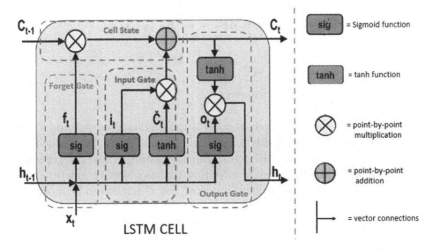

Figure 6.10 A sample LSTM CELL.

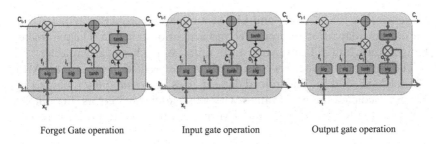

Forget Gate operation   Input gate operation   Output gate operation

Figure 6.11 Three gates operation in LSTM model.

6.7 PERFORMANCE RESULTS

N number of methods is used to evaluate the solar power prediction. The frequently used methods for accurate prediction are:

- Main Bias Error
- Mean Squared Error
- Root Mean Square Error
- Forecast Skill

$$\text{Main Bias Error, MBE} = \frac{1}{s}\sum_{i=1}^{s}\left(O_i - PO_i\right) \tag{6.5}$$

$$\text{Mean Squared Error, MSE} = \frac{1}{s}\sum_{i=1}^{s}(O_i - PO_i)^2 \tag{6.6}$$

$$\text{Root Mean Squared Error, RMSE} = \sqrt{\frac{1}{s}\sum_{i=1}^{s}(O_i - PO_i)} \qquad (6.7)$$

$$\text{Forecast Skill} = 1 - \frac{\text{RMSE}}{\text{RMSE}_{\text{persistence}}} \qquad (6.8)$$

Where O is the actual output and PO is the predicted output and s is the number of samples. The MBE pictures the misjudgment in prediction of solar power. Based on the MBE value, the forecasting can be categorized as underestimated and overestimated. MSE is used as objective function in the training process which has to be minimized. Small errors can be minimized by the value of RMSE.

6.8 RESULTS AND DISCUSSION

Finally, comparing the prediction results of solar forecasting, it gives a convergence that RNN, GRU, and LSTM models are properly predicted with the correct adaptation of optimizer and activation function. The parameters temperature, humidity, pressure, and rainfall were chosen as variables.

The performance metrics considered for the prediction process are MBE, MSE, RMSE, and FS. The trained dataset gives less error and test dataset has almost predicted the prior values with some additional errors. In comparison with the RNN, GRU, and LSTM models, the long-term short memory

Table 6.3 Model parameterization

Training model	Description			
	Batches	Epochs	Optimizer	Activation function
GRU	500	10	ADAM	Tan h
RNN	500	10	ADAM	Tan h
LSTM	500	10	ADAM	Sigmoid and Tan h

Table 6.4 Solar prediction metrics for different models

Training model	Dataset	MBE	MSE	RMSE	FS
GRU	Training	2.63	12.23	3.96	0.76
	Test	2.73	15.48	4.33	0.63
RNN	Training	1.77	7.23	2.68	0.88
	Test	1.86	8.43	2.71	0.96
LSTM	Training	1.63	7.42	2.87	0.93
	Test	1.72	9.21	3.12	0.95

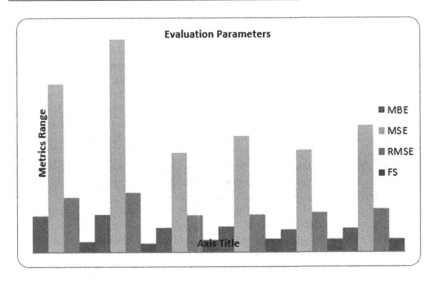

Figure 6.12 Evaluation metrics chart for the models.

trained model able to give reduced error metrics and capable of predicting with the sequential dataset. The GRU and LSTM methods are almost giving nearest results. Depending on the extent of the dataset provided and task performance, any of the one from GRU or LSTM can be chosen for the better model testing.

REFERENCES

[1] Wojtkiewicz, Jessica, Matin Hosseini, Raju Gottumukkala, and Terrence Lynn Chambers. "Hour-ahead solar irradiance forecasting using multivariate gated recurrent units." *Energies* 12, no. 21 (2019): 4055.

[2] Jebli, Imane, Fatima-Zahra Belouadha, and Mohammed Issam Kabbaj. "The forecasting of solar energy based on machine learning." In *2020 International Conference on Electrical and Information Technologies (ICEIT)*, pp. 1–8. IEEE, 2020.

[3] Alzahrani, Ahmad, Pourya Shamsi, Cihan Dagli, and Mehdi Ferdowsi. "Solar irradiance forecasting using deep neural networks." *Procedia Computer Science* 114 (2017): 304–313.

[4] Kingma, Diederik P., and Jimmy Ba. "Adam: A method for stochastic optimization." *arXiv preprint arXiv:1412.6980* (2014).

[5] Anuradha, K., D. Erlapally, G. Karuna, V. Srilakshmi, and K. Adilakshmi. "Analysis of solar power generation forecasting using machine learning techniques." In *E3S Web of Conferences*, Vol. 309, p. 01163. EDP Sciences, 2021.

[6] Zhou, Hangxia, Qian Liu, Ke Yan, and Yang Du. "Deep learning enhanced solar energy forecasting with AI-driven IoT." *Wireless Communications and Mobile Computing* 2021 (2021).

[7] Lynn, Htet Myet, Sung Bum Pan, and Pankoo Kim. "A deep bidirectional GRU network model for biometric electrocardiogram classification based on recurrent neural networks." *IEEE Access* 7 (2019): 145395–145405.

[8] Wang, Huaizhi, Zhenxing Lei, Xian Zhang, Bin Zhou, and Jianchun Peng. "A review of deep learning for renewable energy forecasting." *Energy Conversion and Management* 198 (2019): 111799.

[9] Yadav, Ajay Pratap, Avanish Kumar, and Laxmidhar Behera. "RNN based solar radiation forecasting using adaptive learning rate." In *International Conference on Swarm, Evolutionary, and Memetic Computing*, pp. 442–452. Springer, Cham, 2013.

[10] Che, Zhengping, Sanjay Purushotham, Kyunghyun Cho, David Sontag, and Yan Liu. "Recurrent neural networks for multivariate time series with missing values." *Scientific Reports* 8, no. 1 (2018): 1–12.

[11] Ssekulima, Edward Baleke, Muhammad Bashar Anwar, Amer Al Hinai, and Mohamed Shawky El Moursi. "Wind speed and solar irradiance forecasting techniques for enhanced renewable energy integration with the grid: A review." *IET Renewable Power Generation* 10, no. 7 (2016): 885–989.

[12] David, Mathieu, F. Ramahatana, Pierre-Julien Trombe, and Philippe Lauret. "Probabilistic forecasting of the solar irradiance with recursive ARMA and GARCH models." *Solar Energy* 133 (2016): 55–72.

[13] Azimi, Rasool, Mohadeseh Ghayekhloo, and Mahmoud Ghofrani. "A hybrid method based on a new clustering technique and multilayer perceptron neural networks for hourly solar radiation forecasting." *Energy Conversion and Management* 118 (2016): 331–344.

[14] Batcha, R. Rahin, and M. Kalaiselvi Geetha. "A survey on IOT based on renewable energy for efficient energy conservation using machine learning approaches." In *2020 3rd International Conference on Emerging Technologies in Computer Engineering: Machine Learning and Internet of Things (ICETCE)*, pp. 123–128. IEEE, 2020.

[15] Hache, Emmanuel. "Do renewable energies improve energy security in the long run." *International Economics* 156 (2018): 127–135.

[16] Tripathi, Suman Lata, Mithilesh Kumar Dubey, Vinay Rishiwal, and Sanjeevikumar Padmanaban, eds. *Introduction to AI Techniques for Renewable Energy System*. CRC Press, 2021.

[17] Sandhiya, S. "Experimental analysis on different effects of PV module with respect to panel angle using machine learning algorithms." *Information Technology in Industry* 9, no. 3 (2021): 367–374.

[18] Inbamani, A., P. Umapathy, K. Chinnasamy, V. Veerasamy, and S. V. Kumar. Artificial intelligence and Internet of things for renewable energy systems. *Artificial Intelligence and Internet of Things for Renewable Energy Systems* 12 (2021).

[19] Roshana, A., Abinaya Inbamani, S. Sujeev Rithan, R. Krishnakumar, and J. M. Vishal. "Design and investigation of grid associated PV framework using PVSYST software." In *2021 International Conference on Advancements in Electrical, Electronics, Communication, Computing and Automation (ICAECA)*, pp. 1–6. IEEE, 2021.

[20] Tripathi, Suman Lata, and Sanjeevikumar Padmanabhan. "Green energy: Fundamentals, concepts, and applications". Edited book, Scrivener Publishing, Wiley, 2020. ISBN-10: 1119760763, ISBN-13: 978-1119760764, DOI: 10.1002/9781119760801

[21] Gandhi, R. R., and C. Kathirvel. "A comparative study of different soft computing techniques for hybrid renewable energy systems." In *2021 5th International Conference on Trends in Electronics and Informatics (ICOEI)*, pp. 1667–1677. IEEE, 2021.

[22] Gandhi, R. R., J. A. I. Chellam, T. N. Prabhu, C. Kathirvel, M. Sivaramkrishnan, and M. S. Ramkumar. "Machine learning approaches for smart agriculture." In *2022 6th International Conference on Computing Methodologies and Communication (ICCMC)*, pp. 1054–1058. IEEE, 2022.

Chapter 7

Characterization of CuO–SnO$_2$ composite nano powder by hydrothermal method for solar cell

K. Srinivasan, V. Rukkumani, and V. Radhika

Sri Ramakrishna Engineering College, Coimbatore, Tamil Nadu, India

M. Saravanakumar and S. Kavitha

Gobi Arts & Science College, Gobichettipalayam, Tamil Nadu, India

CONTENTS

7.1 INTRODUCTION

The predominant analysis technique of air quality in the field of auto-motives, spacecrafts, and housing can depend on the materials used in their internal surface. This research focuses on the new materials, and on this technique, to solve problems and increase the efficiency for diode applications. Due to poor dimensions, costs, and power utilization, the metal oxide semiconducting films are the favorable for pn junction diode application.

The tin oxide (SnO$_2$) is the n-type semiconductor [1]. The extreme range of their properties have been employed in electronic devices, optoelectronic applications and Li-ion batteries [2]. One of the conducting oxides of SnO$_2$

DOI: 10.1201/9781003302964-7

transparency and the reflectivity in the visible and infra-red region is high compare with other oxide materials. The resistance of SnO_2 is much smaller than the other oxide materials [3].

The copper oxide (CuO) is a one of the semiconducting materials (p-type) with a band gap range is 1.7–1.2 eV. Hetero junctions can be formed due to doped SnO_2–CuO. The photo electrons can be generated from SnO_2 and migrated to CuO. This is used to separate the photo generated electrons with holes. The methods to be prepare the nano composites CuO–SnO_2 thin films includes liquid phase co-precipitation, thermal evaporation, RF magnetron sputtering, photochemical, and spray pyrolysis methods. In this work, we employ spray pyrolysis techniques to deposit the pure and CuO–SnO_2 thin films.

In this chapter, we introduce different techniques for preparation of different thin film preparation, Section 7.2 gives the sample preparation method, Section 7.3 gives detailed description on results and discussion, Section 7.4 explains about the optical properties, Section 7.5 explains about the diode and solar application, Section 7.6 deals with characteristics of solar cells and finally Section 7.7 concludes.

7.2 SAMPLE PREPARATION METHOD

7.2.1 Preparation of CuO–SnO_2 thin film

Among various techniques of synthesis, we employed the spray pyrolysis method to prepare pure CuO and pure SnO_2 and different ratio of CuO–SnO_2 thin film. The spray solution was prepared by taking copper acetate and tin dihydrate used as copper and tin sources. Copper acetate and tin dihydrate were initially dissolved in 50 ml of deionized water and ethanol to prepare a spray solution. HCL was added to achieve a clear solution. The precursor can be taken in a spray gun fitted with a reservoir. The substrate is cleaned by ultrasonic sonication. In the spray pyrolysis, the temperature can be maintained by the heater, and the pressure is controlled by a pressure meter. The sample is sprayed on the sample surface. Thus, the pure CuO, SnO_2, and CuO–SnO_2 ratio of 0.05(CuO)–0.05 (SnO_2), 0.05(CuO)–0.025 (SnO_2), 0.05(CuO)–0.075 (SnO_2), 0.025(CuO)–0.05 (SnO_2) and 0.075(CuO)–0.05 (SnO_2) were obtained, and they were referred to as C1, S1, CS2, CS5, CS4, CS7, and CS6 respectively. Prepared films were investigated by following characterization techniques.

7.2.2 Characterization techniques

Prepared samples structural properties were carried out by XRD. The surface morphology of the sample were recorded using SEM and HRTEM. The optical properties can be studied by UV absorption studies. The electrical analysis of the CuO–SnO$_2$ thin films were calculated by Keithley 6517-B electrometer in the temperature. Also, the current–voltage characteristics of the CuO–SnO$_2$ junction diode were analyzed using Keithley electrometer.

7.3 RESULTS AND DISCUSSION

7.3.1 Structural properties

The x-ray diffraction pattern of SnO$_2$ and different molar ratio of CuO–SnO$_2$ thin films as shown in Figure 7.1. The XRD peaks of prepared samples were matches JCPDS card no. 45-0937 and 45-1445 respectively. Also, the ratio pattern of CuO–SnO$_2$ matches JCPDS card number of 77-0447. The peaks in the XRD pattern of the films were observed at 2θ value of 28.18, 32.41 and 37.61 which lie in the orientation of (1 1 0), (1 0 1), and (1 1 1) planes respectively.

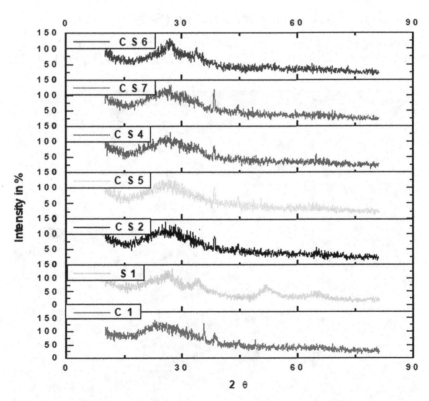

Figure 7.1 XRD pattern of the SnO$_2$ and copper oxide and different ratios of CuO–SnO$_2$ films.

From the Scherer equation, the average crystalline size can be calculated with the following equation.

$$D = 0.89\lambda/\beta \cos\theta \qquad (7.1)$$

where D = Crystalline size, λ = Wavelength for CuKα1 source, β = Full width at half maximum of a diffraction peak and θ = Diffraction angle in degrees.

The grain sizes of the films were calculated in the range 17 to 33 nm. From the result, the CuO ions are well-incorporated with SnO_2 ions due to their atomic radius. From the XRD pattern it is cleared indicated which there formed, SnO_2 as cassiterite tetragonal and for CuO is tenorite mono-clinic [5]. The patterns of the CuO–SnO_2 composite were found to be almost same as those for different ratios, except for CS6 and CS7.

Figure 7.1 clearly shows no additional peaks, indicating that changing ratios do not affect the crystalline structure. However, the intensity of the main prepared nano composite peaks shows the ratio manipulates the transition of amorphous to crystalline state.

7.3.2 Scanning Electron Microscope (SEM)

Morphological structure of pure CuO and Pure SnO_2 and CuO–SnO_2 thin film at different ratios (CuO 0.05, SnO_2 0.025), (CuO 0.05, SnO_2 0.075),

Figure 7.2 SEM images of CuO and SnO_2 thin films.

8/11/2017 9:37:59 AM | HV 20.00 kV | mag ⊞ 30 000 x | det ETD | WD 9.8 mm | spot 3.0 | bias 0 V | ———1 µm———

Figure 7.3 SEM images of CuO and SnO$_2$ thin films.

(CuO 0.05, SnO$_2$ 0.05), (CuO 0.025, SnO$_2$ 0.05), and (CuO 0.075, SnO$_2$ 0.05) with pure CuO and SnO$_2$ as shown in Figures 7.2–7.8

From the SEM analysis, CuO- SnO$_2$ composite thin film with diameter of 1 nm and a uniform array were observed. The shape of the particles suggested that the organic compound used in the synthesis were effectively eliminated during the annealing process. The images verified that the increasing ratios presents to the agglomeration and grain growth leading to a rough particle surface. The SEM images confirmed the grain size calculated from XRD analysis.

7.3.3 Transmission electron microscopy analysis

Figures 7.9 and 7.15 show the TEM micrographs images of pure CuO and SnO$_2$ and CuO–SnO$_2$ thin films prepared at different ratios (CuO 0.05, SnO$_2$ 0.025), (CuO 0.05, SnO$_2$ 0.075), (CuO 0.05, SnO$_2$ 0.05), (CuO 0.025, SnO$_2$ 0.05), and (CuO 0.075, SnO$_2$ 0.05) respectively. Transmission electron

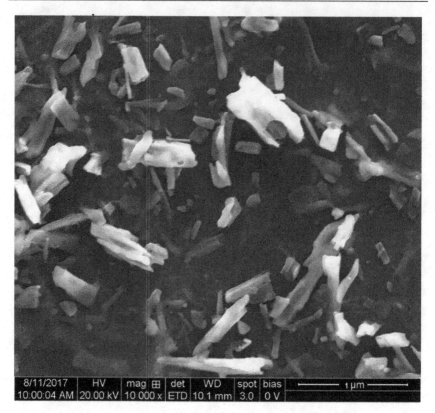

Figure 7.4 SEM images of CuO SnO$_2$ thin film of molar ratio (0.05–0.025).

microscope is used to investigate the micro structure of CuO–SnO$_2$ thin film at different ratios. For transmission electron microscopy analysis, the CuO–SnO$_2$ thin films were dispersed in acetone. The solution was vigorously mixed for 30 minutes.

Figures 7.9–7.15 TEM images of CuO& SnO$_2$ and different molar ratio of CuO SnO$_2$ thin film

The TEM representation of Pure copper oxide (Figure 7.9) shows the flake-like nanostructure is stated in previous work [8]. TEM images (Figures 7.9–7.15) show that CuO–SnO$_2$ composites are well-incorporated. And also confirms the crystalline size which obtained from XRD results. There is an increase in grain size with increase in ratio. As the ratio increases there is increase in oxygen content and this may be due to the solubility of oxygen on increasing the ratios. The variations in the results may be partly due to the presence of strain in the lattice which also contributes to broadening of XRD peaks.

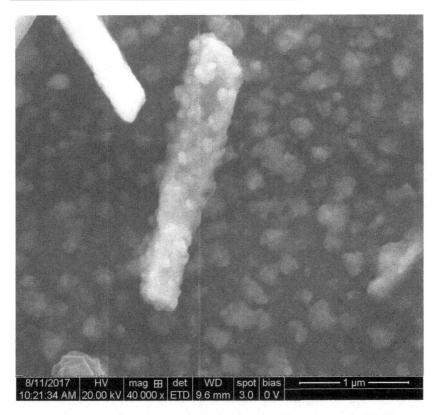

Figure 7.5 SEM images of CuO SnO$_2$ thin film of molar ratio (0.05–0.075).

7.4 OPTICAL PROPERTIES

UV-visible absorption spectroscopy is used for analysis the optical properties of prepared samples. Figure 7.4 shows that the optical absorption spectra of the CuO–SnO$_2$ thin films at different ratios. The energy band gap has been calculated using the Tauc equation

$$E_g = \left(\frac{hc}{\lambda} \right) eV. \tag{7.2}$$

Where h = Planck's constant and c = Velocity of light.

A sharp absorption was observed at around 410 nm for same dopant ratio of CuO and SnO$_2$. The bandgap of the prepared samples lies in the range of 2.98 to 3.67 eV. The prepared samples bandgap slightly decreased due to addition of CuO into SnO$_2$ and is because of the lower band gap of

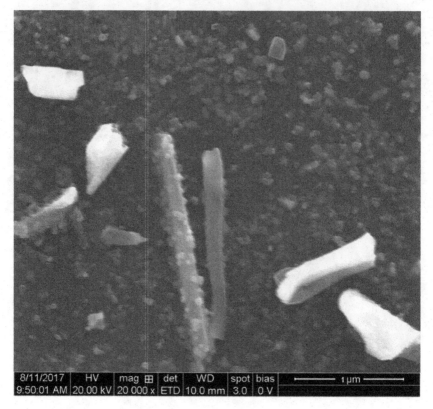

Figure 7.6 SEM images of $CuOSnO_2$ thin film of molar ratio (0.025–0.05).

CuO. When increasing film thickness the band gap is decreased due to the dangling defects in their structure. Furthermore, increasing localized energy states density induce the reducing energy band gap [9].

7.5 DIODE AND SOLAR APPLICATIONS

Due to its promising rectification performance and potential for creating transparent circuits, transparent all oxide-based p-n junction diodes have come under increased scrutiny. The operation of the device is heavily dependent on the quality of the interface between the n- and p-type semiconductor layers, which may be compromised by interfacial surface roughness, third phase impurities, and the formation of interfacial layer [15]. A p-n junction, as its name suggests, is composed of serially connected p- and n-type semiconductors. When choosing materials for n- or p-type transparent semiconducting oxides, interface quality is also important (TSOs).

Figure 7.7 SEM images of CuOSnO$_2$ thin film of molar ratio (0.075–0.05).

When exposed to higher process temperatures for longer processing times, the oxide phase of metals with multiple valence states is more likely to be oxidized or reduced, which would result in the formation of third phase impurities or interfacial layers and reduce the performance of the corresponding device. There are already many n-type TSOs with respectable electrical properties and transparency in the visible region, such as indium oxide (In$_2$O$_3$) [10], zinc oxide (ZnO) [11], and tin dioxide (SnO$_2$).

In comparison to n-type TSOs, the p-type counter parts lag in performance. The most promising p-type oxide semiconductors include ternary copper oxides and nondela fossites(CuSr$_2$O$_2$)], binary copper oxides (CuO and Cu$_2$O), Among these, CuO has been reported with higher p-type Hall mobility values due to the more dispersed valence band maximum, which results from the hybridization between the O$_2$p and Cu$_3$d (or Sn$_5$s) orbitals [12]. However, these two p-type oxides are known to be metastable phases, which have the tendency to be oxidized, forming stable phases with higher metal valence states [16]. The p–n junctions based on these two p-type oxide always suffer from the formation of defective interfacial layers.

| 7/7/2016 | HV | mag □ | det | WD | spot | bias | ————— 1 μm ————— |
| 1:15:25 PM | 10.00 kV | 24 000 x | ETD | 8.7 mm | 3.0 | 0 V | |

Figure 7.8 SEM images of CuOSnO$_2$ thin film of molar ratio (0.05–0.05).

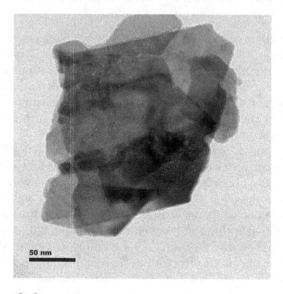

50 nm

Figure 7.9 Pure CuO.

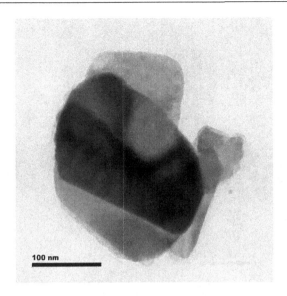

Figure 7.10 Pure SnO$_2$.

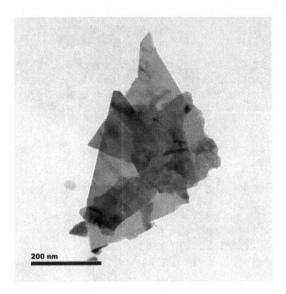

Figure 7.11 CuO (0.05)–SnO$_2$(0.025).

The performance of the p-type counterparts is inferior to that of n-type TSOs. Among these, CuO has been reported to have higher p-type Hall mobility values due to the more dispersed valence band maximum, which results from the hybridization between the O$_2$p and Cu$_3$d (or Sn$_5$s) orbitals. Other promising p-type oxide semiconductors include ternary copper oxides

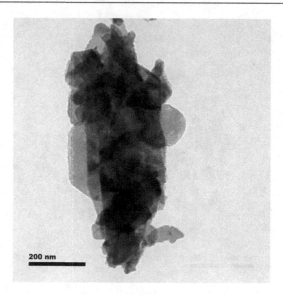

Figure 7.12 CuO(0.05)–SnO$_2$(0.075).

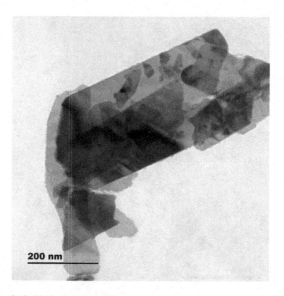

Figure 7.13 CuO(0.025)–SnO$_2$(0.05).

and nondela fossites (CuSr 2 O$_2$)], binary copper oxides (CuO and Cu$_2$O), and ternary copper oxides. However, it is known that these two p-type oxides are metastable phases, which have a propensity to oxidize and transform into stable phases with higher metal valence states [16]. The interfacial layers that develop in p-n junctions based on these two p-type oxides are consistently faulty.

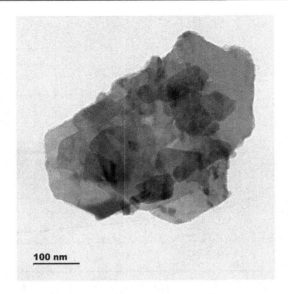

Figure 7.14 CuO(0.075)–SnO$_2$(0.05).

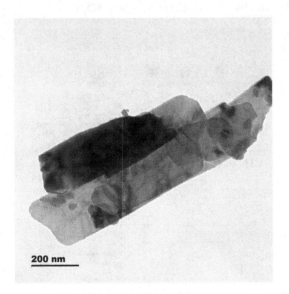

Figure 7.15 CuO(0.05)–SnO$_2$(0.05).

Nowadays, different materials based on SnO$_2$ with CuO as boosting material for diode applications have been prepared and widely studied, such as CuO–SnO$_2$ bi-layers and CuO–SnO$_2$ hetero structures, CuO-doped SnO$_2$ nano ribbons, and CuO-doped SnO$_2$ nano rods. Their diode device is

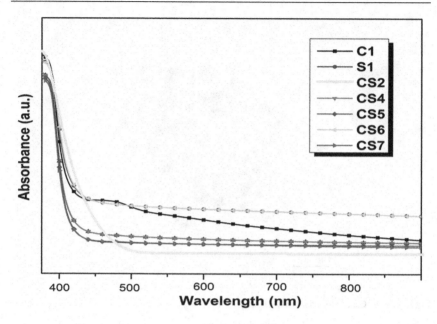

Figure 7.16 UV-vis absorption spectra of $CuOSnO_2$ thin film of different ratios.

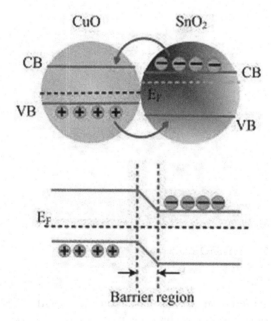

Figure 7.17 CuO–SnO$_2$ in PN junction.

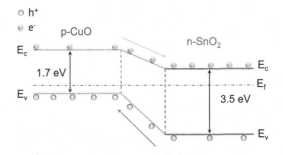

Figure 7.18 p–CuO and n-SnO$_2$ working mechanism.

attributed to the arrangement of p–n junction between n-type SnO$_2$ and p-type CuO, and the powerful resemblance of CuO to diode that tends to the interruption of the p–n junction.

CuO and SnO$_2$ are two standard metal oxide anode materials with theoretical capacity of 670 and 790 mA h g^{-1}, correspondingly, which are approximately two times higher than that of graphite. Additionally, SnO$_2$–CuO composite nanostructures are steady, nontoxic, and environmentally friendly. It has been reported that CuO nano particles deliver a capacity of 470 mA h g^{-1} at a current density of 100 mA g^{-1}[14].And SnO$_2$ nano particles distribute a capacity of 631 mA h g^{-1} at a current density of 1000 mA g^{-1}.

The aim of this research was to optimize the preparation for the CuO–SnO$_2$ nano composite powders from hydrothermal method and explore their diode applications.

7.6 CHARACTERISTICS

The IV characteristics of the CuOSnO$_2$ nano composite thin films at different molar ratios are shown in Figure 7.5. From the graph, linear difference of the current–voltage (I–V) curve shows that the ohmic performance of the prepared nano composite thin films. The electrical conductivity is calculated using the expression of

$$\Sigma = t \, / \, \mathrm{RA}(\Omega \mathrm{cm})^{-1} \tag{7.3}$$

Figure 7.19 shows that the conductivity increases with increasing the ratio of nano composite powders and it reveals that the semiconducting nature of CuO-SnO$_2$ nano composite thin films. The higher conductivity was obtained for equal ratio of CuO-SnO$_2$ nano composite thin films. Figure 7.20 shows the semi-logarithmic plot of ln I-V for CuO-SnO$_2$ nano composite thin films in different ratios. As a homo junction, the p-SnO$_2$–n-CuO diode has prospective advantages over hetero junction diode in affecting a small severe edge imperfection.

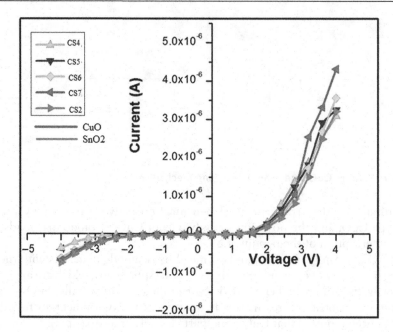

Figure 7.19 IV characteristics of the different molar ratios of CuO–SnO$_2$ thin films.

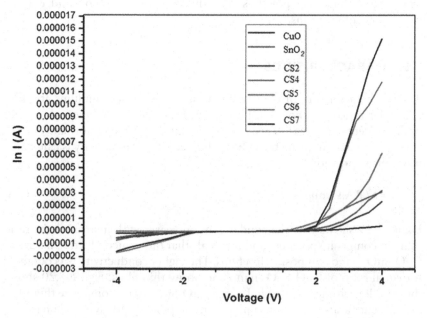

Figure 7.20 IV characteristics of the different ratios of CuO–SnO$_2$ thin films.

7.7 CONCLUSION

The CuOSnO$_2$ nano composite thin films were prepared using the spray pyrolysis method. The prepared samples were characterized by XRD, SEM, TEM UV-Vis spectroscopy and IV characteristics studies. The average grain size of thin films with different ratio from XRD analysis calculated was 17–33 nm. The SEM and TEM images suggested that the which compounds used in the synthesis were effectively eliminated during the annealing process. The absorption wavelength is decreased when increasing dopant molar ratio. The band gaps were 2.99 to 3.67 eV increasingly, depending on their dopant ratio. The higher conductivity was obtained for equal ratio of CuO–SnO$_2$ nano composite thin films. By using the IV studies, the diode properties of the prepared CuO–SnO$_2$ have been analyzed.

REFERENCES

[1] Li, P.G., Guo, X., Wang, X.F., & Tang, W.H. (2009). Synthesis, photo luminesce and dielectric properties of O-deficient Sno$_2$ nanowires. *Journal of Alloys and Compounds*, 479, 74–77.

[2] Wang, C., Li, J., Zhang, Y., Wei, Y., & Liu, J. (2010). Synthesis of Cu-doped Sno$_2$ thin films by spray pyrolysis for gas sensor application. *Journal of Alloys Compound*, 493, 64–69.

[3] Batzill, M., & Diebold, U.(2005).The surface and material science of tin oxide. *Progress in Surface Science*, 79, 47–154.

[4] Sathyaendra, S., Nithi, V., Archana, S., & Yadhav, B.C. (2014). Synthesis and characterization of CuO–SnO$_2$ nano composite and its application as liquefied petroleum gas sensor. *Materials Science in Semiconductor Processing*, 18, 88–96.

[5] Waghuley, S.A., Yenorkar, S.M., Yawale, S.S., & Yawale, S.P. (2007). SnOz/PPy screen-printed multilayer CO$_2$ gas sensor. *Sensors and Transducers Journal*, 79, 1180–1185.

[6] Elangovan, E., Singh, M.P., Dharmaprakash, M.S., & Ramamurthi, K. (2004). Some physical properties of spray deposited SnO$_2$ thin films. *Journal of Optoelectronics and Advanced Materials*, 6, 197–203.

[7] Touidjen, N.H., Bendahmane, B., LamriZeggar, M., Mansour, F., & Aida, M.S. (2016). SnO$_2$ thin film synthesis for organic vapors sensing at ambient temperature. *Sensing and Bio-Sensing Research*, 11, 52–57.

[8] Zou, G., Li, H., Zhang, D., Xiong, K., Dong, C., & Qian, Y.(2006). Precursor-induced hydrothermal synthesis of flowerlike cupped-end micro rod bundles of ZnO. *Journal of Physical Chemistry*, 110, 1361–1363.

[9] Habubi, N.F., Oboudi, S.F., & Chiad, S.S. (2012).Study of some optical properties of mixed SnO$_2$-CuO thin films. *Journal of Nano- and Electronic Physics*, 4, 04008 (4pp).

[10] Lin, Z., Song, W., & Yang, H. (2012). Highly sensitive gas sensor based on coral-like Sno$_2$ Prepared with hydrothermal treatment. *Journal of Sensors and Actuators*, 173, 22–27.

[11] Wu, G., Wang, C., & Zhang, X. (1998). Lithium insertion into CuO/Carbon tubes. *Power Sources*, 75, 175.

[12] Lou, X.W., Wang, Y., & Yuan, C. (2015). Understanding the role of 'path' for sacrificial substance migration during the fabrication of hollow nanostructures in PtPdCu system. *Material Research Express*, 2, 1–10.

[13] Huang, X., Wang, C., & Zhang, S. (2011). CuO/C microspheres as anode materials for lithium-ion batteries. *Electrochimica Acta*, 56, 6752–6756.

[14] Kyaw, K., & Khaing, H.Y. (2020). Fabrication and structural properties of CuO doped Sno_2 nanostructured thin films. *Research Journal*, 11, 169–177.

[15] Sathyaendra, S.,Nithi, V., Archana, S., & Yadhav, B.C. (2014). Synthesis and characterization of $CuO–SnO_2$ nanocomposite and its application as liquefied petroleum gas sensor. *Materials Science in Semiconductor Processing*, 18, 88–96.

[16] Chellaswamy, C., & Ramesh, R. (2016). Parameter extraction of solar cell models based on adaptive differential evolution algorithm. *Renewable Energy*, 97, 823–837.

[17] Supakorn, P., Dheerachai, P., Suttinart, N., Khanidtha, J., & Ki-Seok, A. (2014). Synthesis and characterization of CuO/ SnO_2 nanocomposites. *Applied Mechanics and Materials*, 575, 175–179.

[18] Parthibavarman, M., Hariharan, V., Sekar, C., & Singh, V.N. (2010). Effect of copper on structural, optical and electrochemical properties of SnO_2 nano particles. *Journal of Optoelectronics and Advanced Materials*, 12, 1894–1898.

Chapter 8

Design and development of solar PV based advanced power converter topologies for EV fast charging

S.S. Karthikeyan
Dr. N.G.P. Institute of Technology, Coimbatore, Tamil Nadu, India

R. Uthirasamy
Mahendra Engineering College, Namakkal, Tamil Nadu, India

CONTENTS

8.1 INTRODUCTION TO ELECTRIC VEHICLE CHARGING

Fast charging or rapid charging methods are used to recharge the EV batteries within a minimum time period based on the type of charging methods adopted [1–4]. In fast charging technology, within a minimum period of 20 minutes, 80% of battery packs are charged. The rapid development of the electric vehicle (EV) market has led to the enhancement of new infra for charging stations and charging networks. Slow chargers are typically rated at 3.2 kW, and are used mostly in domestic applications. This type of charger is not practically suitable for other applications, such as EV, testing, arc

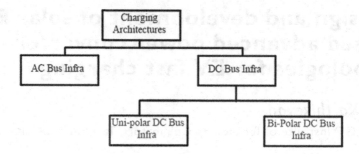

Figure 8.1 Block diagram of charging infra.

welding, etc., because they charge too slowly. Fast chargers are used in car parks and shopping centers, and may charge at rates of 7 kW or 22 kW for EV applications. A 22 kW charger might charge 62 kWh EV from zero to a full charge in approximately two and a half to three hours. Generally, quick chargers frequently charge using DC charging method; inbuilt rectifiers are used in EVs for AC–DC conversion within the charger rather than relying on the EV's onboard rectifier [5–8]. Quick DC chargers are commonly used at up to 52 kW of power. The general block representation of charging infra is shown in Figure 8.1.

At a 52 kW charging rate, our common EV battery is recharged in one hour. EV cars are planned to charge at public charging stations. However, for more convenience, most EV car owners do the greater part of their charging at home. Even though more charging facilities are available in the residential system, consumers are not satisfied with the provided system due to instability in the grid system [9–13]. Generally based on the power rating, charging systems are classified as Level 1, Level 2, and Level 3. A block diagram representation of triple input single output system for EV is shown in Figure 8.2. The equivalent circuit representation of dual output system is shown in Figure 8.3 [14–17].

8.2 DC–DC CONVERTERS

DC–DC converters are widely used for industrial, domestic, and EV applications. In practical consideration, it is necessary for a preset DC input voltage to convert into a changeable DC output voltage. DC–DC converters are generally used for UPS, footing motors in aquatic/marine pumps, elevators, EVs, electric chargers, among other applications. DC–DC converters offer soft acceleration control, high efficiency, and quick response. Much literature has pointed out the remarkable topologies of DC–DC converters. Fundamentally, DC–DC converters are categorized into step-up (boost)

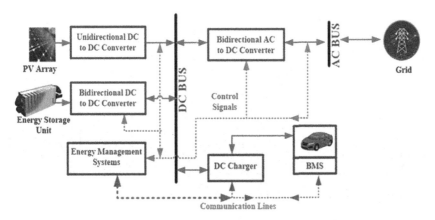

Figure 8.2 Triple input single output system for EV application.

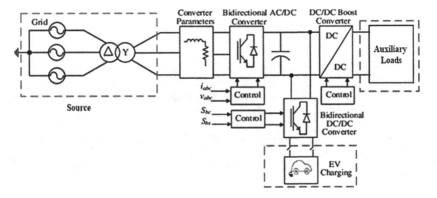

Figure 8.3 Dual output voltage for EV system.

converters, step-down (buck) converters, step-up-step-down (buck-boost) converters, and isolated converters.

8.2.1 Boost converter

It is a step-up converter that is used to boost the level of input voltage in accord with the desired voltage level. The output of step-up converter V_o is always superior to the input voltage V_{dc}.

8.2.2 Buck-boost converter

A buck-boost converter is capable of producing DC output voltage which is either greater or smaller in magnitude than the DC input voltage V_{dc}. The equivalent circuit of the buck converter is shown in Figure 8.6.

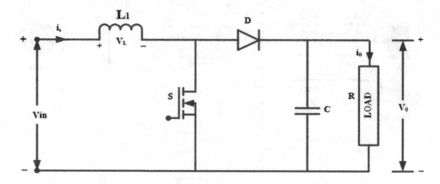

Figure 8.4 Equivalent circuit of boost converter.

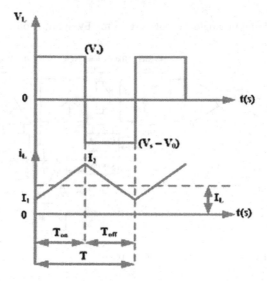

Figure 8.5 Output waveform of boost converter.

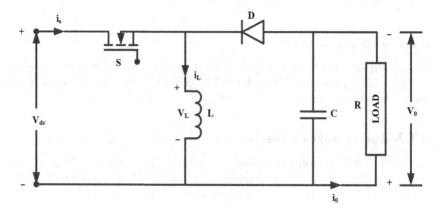

Figure 8.6 Equivalent circuit of the buck-boost converter.

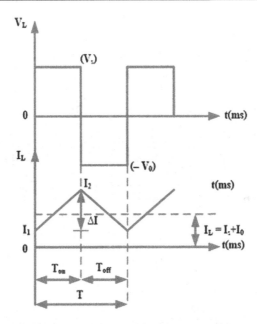

Figure 8.7 Output waveform of buck-boost converter.

Table 8.1 Modes of operation of boost and buck-boost converter

Time period	Control parameters	Boost converter	Buck-boost converter
At $t = T_1$	Voltage across the inductor	$V_L = L_1 \dfrac{I_b - I_a}{T_1}$	$V_L = L \dfrac{I_b - I_a}{T_i}$
At $t = T_1$	Input energy to the inductor	$E_i = V_{in} \cdot I_{s1} \cdot T_1$	$E_i = V_{in} \cdot I_{s1} \cdot T_1$
At $t = T_2$	Output Voltage	$V_O = V_{in} + L \dfrac{dI_{sl}}{T_2}$	$V_O = L \dfrac{I_b - I_a}{T_2}$
At $t = T_2$	Output Voltage	$V_O = \dfrac{V_{in}}{1-D}$	$V_O = \left(\dfrac{D \times V_{dc}}{1-D} \right)$

8.3 DC–DC CONVERTER FOR DC FAST CHARGING

Partial power converter is series-connected with the battery or rectifier unit, where only the difference in voltage between the buses, ΔV, is processed. If the system design is such that the voltage buses are close to each other in value (ΔV small), then the converter processes little power and the losses are minimized. This is in contrast to the full power processing converter, where P_{loss} is a function of the entire battery bus voltage. With sufficiently low ΔV, P_{loss} decreases, and the efficiency increases as a result. If ΔV max is sufficiently small, the converter power rating becomes small. The total power transfer from input to output, however, is not affected. DC fast charging system for EVs is represented in Figure 8.8.

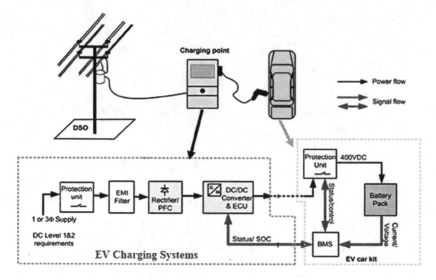

Figure 8.8 DC fast charging system for EVs.

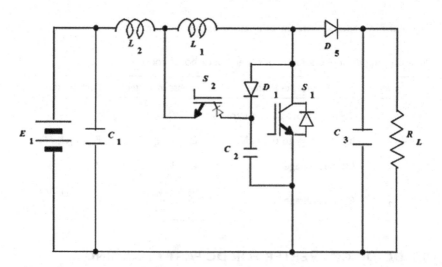

Figure 8.9 Equivalent circuit of SAZZ converter.

8.3.1 Snubber assist zero-voltage zero-current (SAZZ) converter for EV applications

Normally, SAZZ (snubber assist zero-voltage zero-current) converter is used for the Fuel Cell interfaced Electric Vehicle (FCEV) or the Hybrid vehicle applications. The SAZZ pattern is operated like that of the resonant converter. Additionally, the system is designed in such a way to achieve zero-voltage spikes during the diode operation. Equivalent circuit of SAZZ converter is shown in Figure 8.9.

8.3.1.1 Modes of operation

SAZZ configuration has the operating structure of ZVT and ZCT during ON state and ZVS during OFF state. In general voltage and current commutation process, capacitor C_2 is designed to discharge through auxiliary switch S_2 to trigger ON state of the main switch S_1 which reduces the voltage and current stress across the main S_1 through the supplementary control S_2. The main switch S_1 is triggered to OFF state, and the voltage across switch S_1 begins to ascend as of zero potential by the capacitor C_2. Consequently, it can be operated to achieve the quasi-resonant operation which ensues in the exclusion of reactor S_{L1}. When the switch S is anticipated in the designed manner, the proposed converter has the improvements from the quasi-resonant operation scheme.

The modes of operation of the proposed converter are explained in Table 8.2:

Table 8.2 The modes of operation of the proposed converter

Modes of operation	Components	Working
Mode I	S_1	OFF State
At t = −t_2	S_2	ON State
	D_1	OFF State
	D_5	ON State
	C_2	Discharging
	C_3	Charging
	L_1	Energizing
	L_2	Energizing
Mode II	S_1	ON State
At t = −t_1	S_2	OFF State
	D_1	ON State
	D_5	OFF State
	C_2	Charging
	C_3	Charging
	L_1	De-energizing
	L_2	De-energizing
Mode III	S_1	ON State
At t = t_0	S_2	ON State
	D_1	ON State
	D_5	OFF State
	C_2	Discharging
	C_3	Charging
	L_1	Energizing
	L_2	Energizing

(Continued)

Table 8.2 (Continued)

Modes of operation	Components	Working
Mode IV	**S₁**	ON State
At t = t₁	**S₂**	OFF state
	D₁	OFF state
	D₅	OFF state
	C₂	Charging
	C₃	Charging
	L₁	Energizing
	L₂	Energizing
Mode V	**S₁**	OFF state
At t = t₂	**S₂**	OFF state
	D₁	ON state
	D₅	OFF state
	C₂	Discharging
	C₃	Charging
	L₁	De-energizing
	L₂	De-energizing
Mode VI	**S₁**	OFF State
	S₂	OFF state
	D₁	OFF state
	D₅	OFF state
	C₂	Charging
	C₃	Charging
	L₁	De-energizing
	L₂	De-energizing

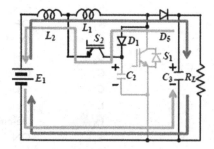

Figure 8.10 Mode I operation of SAZZ converter.

8.3.2 Re-lift converter

A re-lift converter operates based on the voltage lifting technique. It can give the output voltage equal to two times the input voltage. The re-lift converter consists of two power semiconductor switches, S_1 and S_2, three

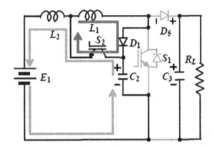

Figure 8.11 Mode 2 operation of SAZZ converter.

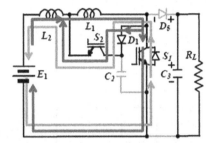

Figure 8.12 Mode 3 operation of SAZZ converter.

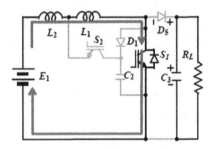

Figure 8.13 Mode 4 operation of SAZZ converter.

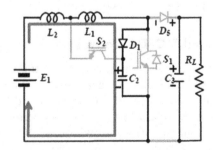

Figure 8.14 Mode 5 operation of SAZZ converter.

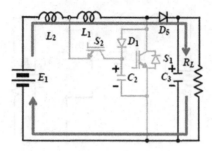

Figure 8.15 Mode 6 operation of SAZZ converter.

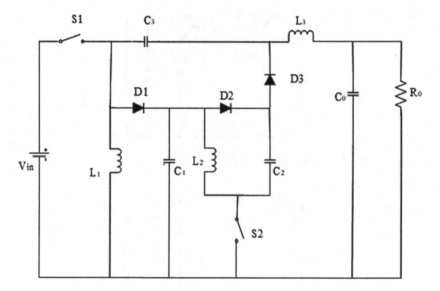

Figure 8.16 Circuit diagram of Re-lift converter.

energy transferring inductors, L_1, L_2, and L_3, four charge transferring capacitors, C_1, C_2, and C_3, and C_4, and three unidirectional devices called diodes, D_1, D_2, and D_3. The proposed re-lift converter is represented in Figure 8.16.

8.3.3 Bi-directional DC–DC converter

The proposed converter system is designed to attain both the buck and boost operation for EVs with a built regenerative braking system. Whenever the battery is charging, the proposed converter should be operated in buck mode and throughout the discharging, it should operate in the boost mode. When the converter consists of both low voltage V_L and high voltage V_H sides, The DC bus is connected in the V_H region and the battery source is connected to the V_L side of the network. The proposed circuit comprises two switches named S_1 and S_2. However, this is a trouble-free in configuration, and it is not appropriate for a large choice of applications because the boost voltage gain is incomplete because of the consequence of the power semiconductor switches. Thus, the predictable converter is not appropriate for a wide voltage-conversion application. The proposed converter has the intrinsic worth of easy configuration and controllability.

The operation of the bidirectional converter consists of four semiconductor switches, named S_1, S_2, S_3, and S_4. It can be operated in two operating modes, buck mode and boost mode. Each mode has three stages of operations. During the buck mode of operation, the HV side DC bus or the solar PV acts as a source and the battery acts as the load. The battery life is affected by more reasons and one of them is overcharging of the battery. So, the battery should be charged only at low voltage. So the bidirectional converter acts as a buck converter during charging. The modes of operation are as follows:

In Stage 1, CCM operation can be obtained in the time period of $(t_0 - t_1)$. In the period in between, S_3 is designed in such a way to operate as synchronous AC–DC converter.

The current through the inductor L_1 is obtained as

$$I_{L1(\text{STEP-DOWN})} = I_{L1}(t_0) + (1/L_1) \times ((V_H/2) - V_L) \times (t - t_0)$$

In Stage 2 operation, it can be operated in the time period of $[t_1-t_2]$. The current through the inductor L_1 is obtained as

$$I_{L1(\text{STEP-DOWN})} = I_{L1}(t_1) \times (V_L/L_1) \times (t_1 - t_0)$$

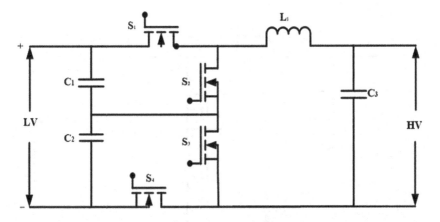

Figure 8.17 Circuit diagram of bidirectional DC–DC converter.

Table 8.3 Various operational stages of buck mode

Stage	ON state switches	OFF state switches	The voltage across the inductor
1	S_1 and S_3	S_2 and S_4	$V_{L1(STEP\text{-}DOWN)} = (V_H / 2) - V_L$
2	S_2 and S_3	S_1 and S_4	$V_{L1\,(STEP\text{-}DOWN)} = -V_L$
3	S_2 and S_4	S_1 and S_3	$V_{L1\,(STEP\text{-}DOWN)} = (V_H / 2) - V_L$

In Stage 3 converter operated in buck, the mode can be derived the time interval of $[t_2 - t_3]$. Further, the current through the inductor L_1 is obtained as

$$I_{L1(STEP\text{-}DOWN)} = I_{L1}(t_0) + (1 / L_1) \times ((V_H / 2) - V_L) \times (t - t_2)$$

8.3.4 Boost mode or step-up mode of bidirectional DC–DC converter

Whenever the power from the solar PV is not sufficient to the load, the discharging process of battery would take place. Normally, the output voltage of battery is very low, at around 12 V. So, it is needed to boost the output voltage from the battery for satisfying the maximum load. During this condition, the bidirectional converter is intended to function in step-up mode or boost mode.

During the boost mode operation, the battery source operated as a supply for high voltage DC bus or load. Various operational stages of boost mode are as follows:

The Stage 1 operation can take place in the time interval of (t_0-t_1). The energy of the low voltage side V_L is transferred to the inductor L_1. The

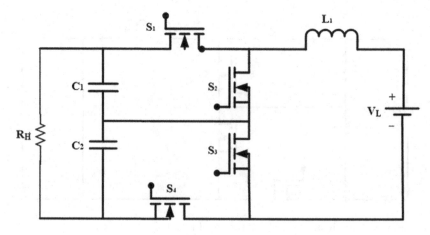

Figure 8.18 Boost mode operation in bidirectional converter.

capacitors S_1 and S_2 are loaded up to discharge for the load RH. The current through the inductor L_1 is obtained as

$$I_{L1(STEP-UP)} = I_{L1}(t_0) + (V_L / L_1) \times (t - t_0)$$

In Stage 2 operation, the step-up mode can take place in the time interval of (t_1-t_2). The energies of the low voltage side VL and inductor L_1 are in series to release their energy to the capacitor C_1. The capacitors C_1 and C_2 are stacked to discharge to the load RH. The current through the inductor L_1 is given as

$$I_{L1(STEP-UP)} = I_{L1}(t_1) + (V_L / L_1) \times (t - t_1)$$

The Stage 3 operation of step-up mode can take place in the time interval of (t_2-t_3). The operation principle is the same as the Stage 1. The current through the inductor L_1 is obtained as

$$I_{L1(STEP-UP)} = I_{L1}(t_2) + (V_L / L_1) \times (t - t_2)$$

Table 8.4 Various operational stages of boost mode

Stage	ON state switches	OFF state switches	Voltage across the inductor
1	S_2 and S_3	S_1 and S_4	$V_{LI(STEP-UP)} = V_L$
2	S_1 and S_3	S_2 and S_4	$V_{LI(STEP-UP)} = V_L / (V_H/2)$
3	S_1 and S_2	S_3 and S_3	$V_{LI(STEP-UP)} = V_L$

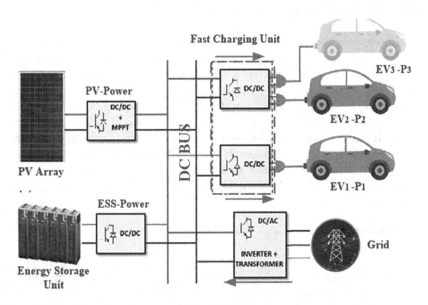

Figure 8.19 Bidirectional converter-based EV charging.

Table 8.5 Comparative analysis of DC–DC converters

Parameter	Conventional converters	SAZZ	Re-Lift	Bidirectional
Device Stress	High	Very Low	Low	High
Operation	Unidirectional	Unidirectional	Unidirectional	Bidirectional
Conduction Loss	High	Low	Low	Medium
Turn ON switching loss	Low	Very Low	Low	Low
Turn OFF switching loss	High	Very Low	Low	High
Switching Frequency	Medium	High	High	High
Efficiency	Medium	High	High	Medium
Boost Factor	Medium	High	High	High

The comparative analysis of various DC–DC converters is represented in the above table. From the analysis, it is inferred that each and every converter has its own merits and demerits. Based on the application of the particular converter, the merits of the owner have been highlighted. Voltage stress is zero in the SAZZ converter, but on the other side, the bidirectional property has been restricted. Likewise, re-lift converters are perfectly designed for EV fast charging and the bidirectional converter is designed and implemented for EV applications.

8.4 CONCLUSION

In this chapter, several conventional and advanced DC–DC converters have been reviewed and discussed, and further inferences have been provided for different converters' EV battery charging operations. Advanced and promising bidirectional DC–DC converters, such as snubber assist zero-voltage zero-current (SAZZ) converter and re-lift converter, have been introduced, along with their topologies and control strategies for EV fast charging operation. It flows power in a bidirectional way between vehicle-to-grid and grid-to-vehicle, and should be such that it reduces the impacts in power grid in terms of load levelling, peak load shaving, and improving voltage regulation and stability of the power grid station.

REFERENCES

1 Hemavathi S, "Chapter 4: Modeling and Energy Optimization of Hybrid Energy Storage System", in *Hybrid Renewable Energy Systems*, Book Series: Advances in Renewable Energy, pp. 97–114 (2021). https://doi.org/10.1002/9781119555667.ch4

2   Hemavathi S, "Modeling of Analog Battery Management System for Single-cell Lithium-ion Battery", *Energy Storage*, Vol. 3, Issue 4, pp. 1–9 (2020). https://doi.org/10.1002/est2.208

3   Hemavathi S, "Overview of Cell Balancing Methods for Li-ion Battery Technology", *Energy Storage*, Vol. 3, Issue 2, pp. 1–12 ([2020] 2021). https://doi.org/10.1002/est2.203

4   Hemavathi S, "Chapter 17: Li-ion Battery Health Estimation based on Battery Internal Impedance Measurement", in *Innovations in Sustainable Energy and Technology*, Book Series: Advances in Sustainability Science and Technology, 1st ed., pp. 183–193 (2021). https://doi.org/10.1007/978-981-16-1119-3_17

5   Agamy MS, Chi S, Elasser A, Harfman-Todorovic M, Jiang Y, Mueller F & Tao F, "A High-power-Density DC–DC Converter for Distributed PV Architectures", *IEEE Journal of Photovoltaics*, Vol. 3, Issue 2, pp. 791–798 (2012). https://doi.org/10.1109/PVSC-Vol2.2012.6656721

6   Hemavathi S, "Chapter 12: Modeling and Estimation of Lithium-ion Battery State of Charge Using Intelligent Techniques", in *Advances in Power and Control Engineering*, Book Series: Lecture Notes in Electrical Engineering, Vol. 609, pp 157–172 (2019). https://doi.org/10.1007/978-981-15-0313-9_12

7   Hemavathi S & Singh M, "Microgrid Short Circuit Studies", *IEEE Xplore*, pp. 1–6 (2019). https://doi.org/10.1109/POWERI.2018.8704389

8   Bellur DM & Kazimierczuk MK, "DC–DC Converters for Electric Vehicle Applications", *Electrical Insulation Conference and Electrical Manufacturing Expo*, pp. 286–293 ([2007] 2008). https://doi.org/10.1109/EEIC.2007.4562633

9   Chen G, Lee Y-S, Hui S, Xu D & Wang Y, '"Actively Clamped Bidirectional Flyback Converter", *IEEE Transactions on Industrial Electronics*, Vol. 47, Issue 4, pp. 770–779 (2000). https://doi.org/10.1109/41.857957

10 Chen L, Wu H, Xu P, Hu H & Wan C, "A High Step-down Non-isolated Bus Converter with Partial Power Conversion Based on Synchronous LLC Resonant Converter", in *IEEE Applied Power Electronics Conference and Exposition (APEC)*, pp. 1950–1955 (2015). https://doi.org/10.1109/APEC.2015.7104614

11 Duan R-Y & Lee J-D, "High-efficiency Bidirectional DC–DC Converter with Coupled Inductor", *IET Power Electronics*, Vol. 5, Issue 1, pp. 115–123 (2012). https://doi.org/10.1049/iet-pel.2010.0401

12 Ghahderijani MM, Castilla M, Momeneh A, Miret JT & de Vicuna LG, "Frequency-modulation Control of a DC/DC Current-source Parallel-resonant Converter", *IEEE Transactions on Industrial Electronics*, Vol. 64, Issue 7, pp. 5392–5402 (2017). https://doi.org/10.1109/TIE.2017.2677321

13 Kimball JW & Krein PT, "Singular Perturbation Theory for DC–DC Converters and Application to PFC Converters", *IEEE Transactions on Power Electronics*, Vol. 23, Issue 6, pp. 2970–2981 ([2007] 2008). https://doi.org/10.1109/PESC.2007.4342105

14 Konjedic T, Korošec L, Truntic M, Restrepo C, Rodic M & Milanovic M, "DCM-based Zero-Voltage Switching Control of a Bidirectional DC–DC Converter with the Variable Switching Frequency", *IEEE Transactions on Power Electronics*, Vol. 31, Issue 4, pp. 3273–3288 (2015). https://doi.org/10.1109/TPEL.2015.2449322

15 Morrison A, Zapata JW, Kouro S, Perez MA, Meynard TA & Renaudineau H, "Partial Power DC–DC Converter for Photovoltaic Two-stage String Inverters", in *IEEE Energy Conversion Congress and Exposition (ECCE)*, pp. 1–6 ([2016] 2017). https://doi.org/10.1109/ECCE.2016.7855332

16 Kwon M, Oh S & Choi S, "High Gain Soft-switched Bidirectional DCDC Converter for Eco-friendly Vehicles", *IEEE Transactions on Power Electronics*, Vol. 29, Issue 4, pp. 1659–1666 (2013). https://doi.org/10.1109/TPEL.2013.2271328

17 Hemavathi S & Nithiyananthan K, "DC to DC Energy Conversion using Novel Loaded Resonant Converter" *International Journal of Power Control and Computation (IJPCSC)*, Vol. 7. Issue 2, pp. 113–122 ([2013] 2015). http://ijsetr.com/uploads/146523IJSETR340-08.pdf

Chapter 9

Assessment of different MPPT techniques for PV system

M. Karthik and N. Divya

Sri Ramakrishna Engineering college, Coimbatore, Tamil Nadu, India

CONTENTS

9.1 INTRODUCTION

Among the sources of environmentally friendly power are sunlight-based energy. Rather than gasoline and other customary exhaustible powers (e.g., coal), solar energy is unadulterated, boundless, and without cost [1]. PV producing frameworks, however, have two significant issues: the adequacy of electric power change. The volume of electric power delivered by a sun-oriented module constantly changes with environment conditions [2]. Moreover, the VI sun-powered cell has a nonlinear property that varies with radiation and temperature. As a general rule, the Maximum Power Point technique (MPP), where the whole PV framework is most effective, and creates its optimal power is the key factor on the current vs. voltage VI, or voltage vs. power VP bend. At its best, the MPP's position is obscure, it tends to be tracked down utilizing search calculations or ascertaining models. The working place of the PV module is kept at the Maximum Power Point (MPP) utilizing Maximum Power Point and Tracking (MPPT) systems [3]. Various MPPT techniques, including the Incremental Conductance approach, Artificial Neural Network framework, the perturb and observe framework, Fuzzy framework, and so forth, have been proposed. Nonetheless, following the MPP on the PV or IV bend has not yet been finished, and standard P&O has constraints including slow following rate that doesn't follow the exact most extreme power point during unexpected changes in light and temperature. Thus, this study alters the customary P&O approach and uses MATLAB programming and Simulink results to assess the INC, P&O, and MP&O strategies [4].

9.2 MODELING OF PV PANEL

Since the photovoltaic cell is the central part of a PV framework, precise displaying of a photovoltaic cell is essential for building a compelling PV framework [5]. Various examinations have recently centered around demonstrating PV modules as well as topological portrayals that are used either alone or within a network. The geography framework that is utilized is essential for fruitful PV module displaying. There are a few numerical models of PV cells that can be tracked down in the writing, including the ideal perfect model, the two-terminology diode model, and the single-point diode model [6]. A photograph produced current source (I_{ph}) and a diode can be aligned to address an optimal PV module, as per the laws of physical science, shown in Figure 9.1a. The single-point diode D interprets the PV module's p-point intersection, and the ongoing moving through it, Id, portrays the breaks through the p–n intersection resulting from the dispersion instrument. The easiest model is one that is accepted to be loss-less [7]. Be that as it may, this model is definitely not a genuine portrayal of a PV module's construction. As outlined in Figure 9.1b, the PV module's series obstruction R_s, which represents the conductance misfortune, is considered to increment exactness. The spillage current in the p–n intersection is addressed by another opposition, R_{sh}, that is mentioned in Figure 9.1c, to develop precision. The development has changed to incorporate a subsequent diode. Figure 9.1c as found in Figure 9.1, the better model, is known as a two-diode model, and was made to develop display clarity (d). The dispersion current brought about by critical charges is addressed by current Id1 through diode D1 in this model, while the recombination current brought about by minor charges is addressed by current Id2 through diode D2. Also, a two-diode model's behaviour generally looks like that of a real PV module, the model is intricate and non-direct. Its numerical investigation is incredibly difficult [8]. Despite being non-direct, the single-diode PV module model has a less difficult construction than the two-diode model. Thus, this model's investigation is less difficult than the two-terminology diode model. It re-joins rapidly to any adjustments in the framework's circumstances [9]. Looking at the different PV module models that have been accounted for, it can be seen that the single-diode-five-boundary model, which utilizes five boundaries, including R_s series method of obstruction, R_{sh} – shunt point opposition, single-point diode ideality factor (a), dull immersion current, and I_{ph} – photograph produced

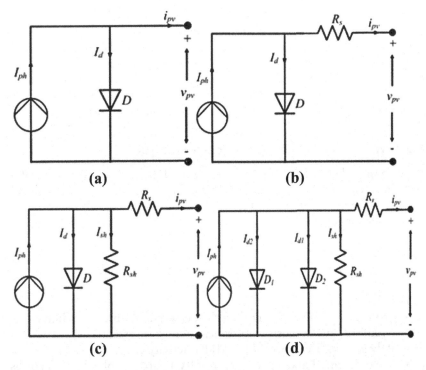

Figure 9.1 (a) Ideal diode type; (b) One diode poly parameter type; (c) One diode five border parameter type; and (d) double-point diode type of a PV unit.

current is reasonable in keeping a streamlined harmony between impersonations of the actual PV module and the effortlessness of execution in numerical examination. Subsequently, in this work, a solitary diode five-boundary model is considered.

9.3 MPPT PROCEDURES

The greatest power point of a PV module would rely upon the encompassing temperature and sunlight-based radiation. Optimal power would be provided to the heap in the event that a heap line crossed at this point, and most extreme power point changes with varieties in temperature or potentially insolation [10]. The level of power can't be conveyed to the heap in light of the fact that the heap line is consistent and doesn't go through the greatest power point. The genuine burden line point should be meant the most extreme power point by adjusting the obligation pattern of the DC–DC converter to guarantee the exchange of greatest power, which expects that the heap follow the greatest power point [11]. To run the PV framework at its pinnacle power, we can physically change the requirement sequence (D) of the Direct Current converter (V_{mpp}, I_{mpp}). Following the most extreme power source point as the temperature and occurrence sun-oriented radiation vary over the course of the day, obligation cycle (D) can be modified to work on a programmed premise. There are various techniques that can be utilized in simple or computerized strategies to naturally change obligation cycle (D).

9.3.1 Incremental conductance technique

At the point when MPP is reached, the PV module keeps on working, unless an adjustment of dIPV is seen, the calculation changes the voltage of the PV exhibit (V PV) will follow another MPP. The addition size controls the following velocity of the MPP.

Advantages:
- Even with frequently moving meteorological circumstances, great yield

Disadvantages:
- Proficiency is much lower than the perturb and observe technique
- Muddled and expensive controller circuits are essential
- Poly sensors are expected for MPPT activity
- Yield potential and current signs of PV boards wobble level in consistent mode

9.3.2 Perturb and observe procedure

The most common approach is the slope climbing (perturb and observe) calculation [12]. The convenience gives it its ubiquity. It has gone through significant exploration, and there are various varieties with little contrasts. The method involved with getting to the greatest power point is iterative. The working point is disturbed, trailed by the estimation of the framework's reaction to distinguish the course of the following irritation, which builds the PV voltage while expanding the PV yield power on the left node of the MPP. Arranged on the other hand, the right node of the bend shows that power develops as voltage diminishes. As needs be, in the event that the power increments after a bother, the resulting irritation will continue in a similar heading. The bearing is switched, assuming that the power brings down. The PO – perturb and observe – strategy is important for the slope climbing approach. Figure 9.3 shows a stream diagram of this strategy. The base voltage $V_{ref}(k)$ for the PV display voltage regulator is resolved utilizing the calculation. The quick voltage and current, V(k) and I(k), are utilized to decide the power right now, P(k). Following P (k), the force of occurrence P is analyzed (k1). The program next assesses the latest modification in the PV value exhibits potential change and endures to alter it in a similar course,

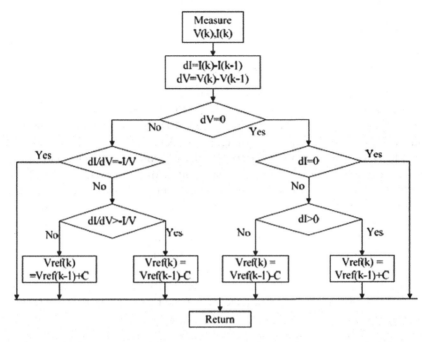

Figure 9.2 Flowchart of steady conductance MPPT strategy.

either by adding or deducting gradual worth C from the reference voltage, in the event that the power value enlarges. Assuming the power has diminished, then the adjustment of voltage is designed to go the alternate way. The framework sways near the MPP in the wake of rehashing this cycle until it accomplishes the MPP. The size of the annoyance and update recurrence both impact the swaying's size. It is feasible to work on the calculation to diminish wavering. This technique's inability to work in conditions with quickly differing irradiance is one of its drawbacks. This happens while the working point changes in the other course on the grounds that the power change created by the climatic condition is more prominent and the other way than the progressions brought about by the calculation's bother.

Advantages:
- Precise result
- Successful and trustworthy strategy
- Unaffected by the board's credits and properties

Disadvantages:
- Size of bother influences exactness and required time
- Inappropriate for quickly changing ecological circumstances
- Indeed, even at consistent express, the result voltage and current signs of PV boards wobble

Consequently, we pick the changed perturb and observe Method to get rid of the deficiencies of these two strategies.

9.3.3 Altered P&O process

By wiping out the erroneous control peculiarity during a period when irradiance is quickly changing, the changed P&O technique is carried out in view of the customary P&O MPPT strategy [13, 14]. As portrayed in Figure 9.5. Subsequent to seeing the PV voltage, the P&O technique's idea makes it simple to look at the power variety and the PV voltage reference. Generally, it is predicated on the idea that the main source of force variety is an adjustment of PV voltage [15, 16]. Essentially, the output power variety of a PV exhibit could create by changes in irradiance vacillation and the guideline of the PV inverter, among other ecological variables [17–19]. At the point when the irradiance circumstance in Figure 9.1 quickly changes from irradiance (i) at t1 to irradiance (ii) at (t+δt), as standard P&O strategy can't be applied in Figure 9.4, the PV inverter could glitch.

The PV inverter explicitly trains the value of PV voltage to be raised from V_1 to next level V_2 under irradiance (i) at time t_1, accepting that the former

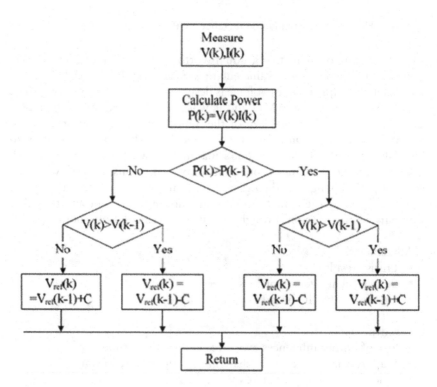

Figure 9.3 Process flow of slope method of climbing PO calculation.

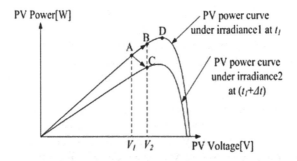

Figure 9.4 Malfunction peculiarity of the traditional MPPT method of control when the illuminance is changing from irradiance (i) to irradiation (ii).

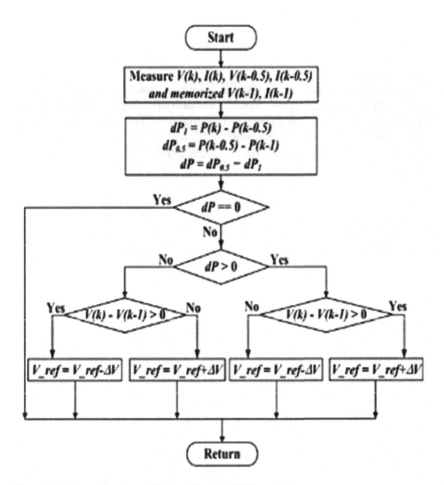

Figure 9.5 Flow diagrams of the adjusted P&O technique.

MPPT order taught the PV voltage to be raised. The irradiance changes rapidly, subsequently after a short timeframe t, the following functional value stays at point value C as opposed to point value B in Figure 9.4. The MPPT regulator for PV type of inverter can screen the right bearing to distinguish the greatest power point of PV module in light of the power variety with just MPPT order in (4).

This peculiarity is best portrayed as a fall in PV power following an expansion in PV voltage order. Thus, the regular P&O calculation would diminish the accompanying PV voltage order. The method is to follow the original greatest power point (MPP), which is point value of D in Figure 9.4, the other way. To this end, under quickly evolving irradiance, the customary P&O approach can't follow MPP. The changed P&O strategy is presented as an answer for this irradiance unsettling influence to recognize the power variety welcomed on by an irradiance shift and MPPT control. The better P&O approach, as found in Figure 9.5, adds every second estimation of PV method exhibit power at the midpoint of the MPPT control unit period.

To diminish the effect of commotion, normal PV still up in the air for MPPT control. The power differential dP 0.5 in (2) between the MPPT control's starting power P (k-1) and midpoint power P (k-0.5) incorporates both the power change brought about by the MPPT control and the irradiance change. In any case, just the power welcomed on by an irradiance shift is addressed by the worth dP1 in (3). A power distinction dP prompted by the sole MPPT control order can be assessed because of (2) through (4).

Advantages:
- Results are precise
- Following is speedy
- Effectiveness is higher than with the P&O and Inc strategies

9.4 RESULTS AND DISCUSSIONS

Case 1: Different kind of illuminations and steady temperature of 25° C.

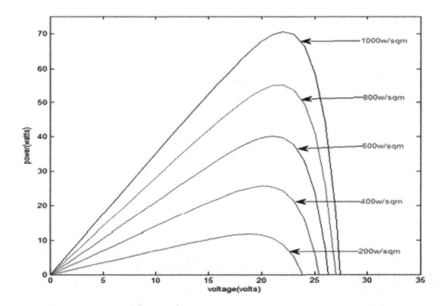

Figure 9.6 Power – voltage characteristics.

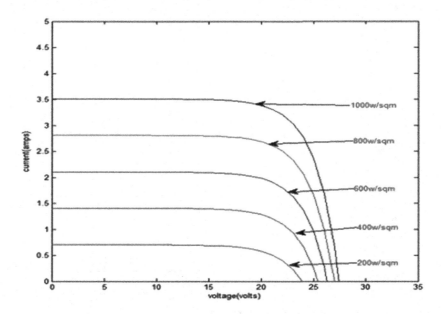

Figure 9.7 Current – voltage characteristics.

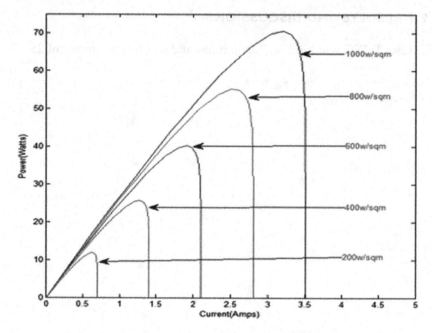

Figure 9.8 Power – current characteristics.

Case 2: Different temperatures and steady light of 1000 w/sqm

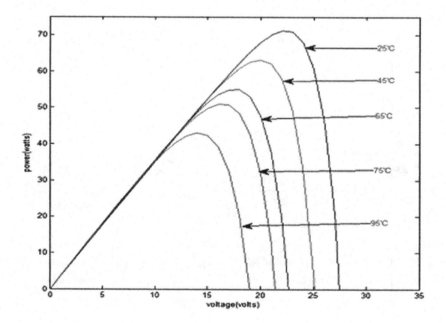

Figure 9.9 Power – voltage characteristics.

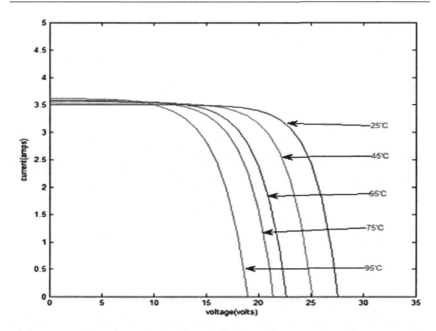

Figure 9.10 Current – voltage characteristics.

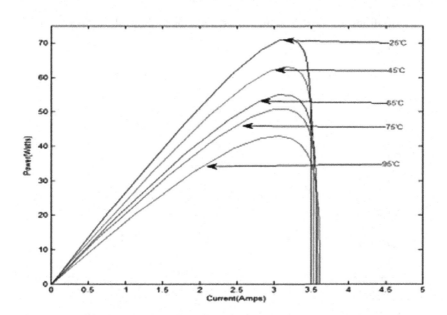

Figure 9.11 Power – current characteristics.

Following of Maximum Power Point PV qualities for various illumination levels at steady temperature of 25° C.

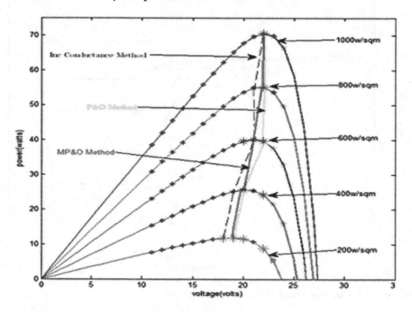

Figure 9.12 Tracking of MPPT using conductance method, perturb and observe, and advanced PO methods for different irradiation levels and constant temperature i.e, 25° C.

PV qualities for various temperatures at steady light of 1000 w/sqm.

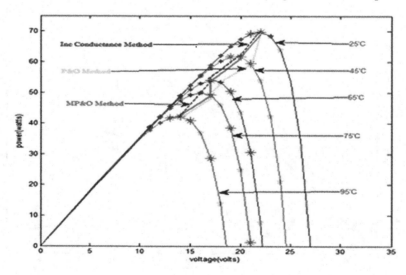

Figure 9.13 Tracking of MPPT using conductance method, Perturb observe and advanced PO methods for different temperature levels and constant irradiation i.e., 1000 w/sqm.

Correlation of INC, P&O and MP&O – MPPT Techniques

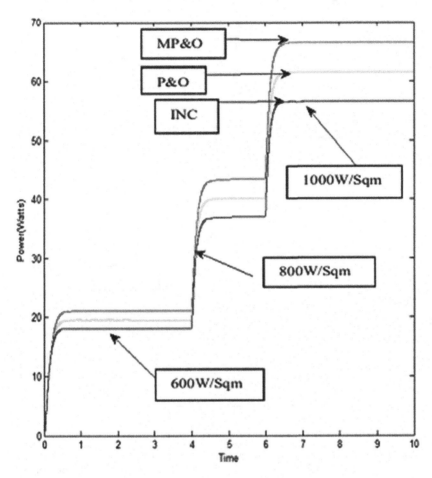

Figure 9.14 Tracking of MPPT using conductance method, Perturb observe and advanced PO methods for different irradiation levels and constant temperature i.e, 25° C and a particular load.

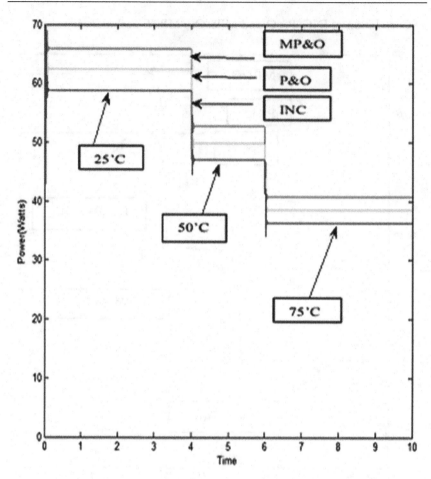

Figure 9.15 Tracking of MPPT using conductance method, Perturb observe and advanced PO methods for different temperature levels and constant irradiation i.e, 1000 w/sqm.

Table 9.1 Time evaluation of conductance method, PO and MPO

Technique	Time to track MPP for different irradiations(sec)	Time to track MPP for different temperatures(sec)
INC	14.917	17.065
P&O	14.012	15.967
MP&O	2.947	4.631

Table 9.2 Power assessment of conductance method, PO and MPO procedures for diverse irradiations and Temperatures

Technique	Power for different irradiations			Power for different temperatures		
	1000	800	600	25	50	75
INC	56.55	36.99	18.18	58.82	47.04	36.34
P&O	61.46	40.22	19.57	62.38	49.89	38.54
MP&O	66.37	43.44	21.14	65.95	52.74	40.75

Table 9.3 Efficiency assessment of conductance method, PO and MPO procedure

Technique	Conversion efficiency (%) at irradiation of 1000 W/sqm	Conversion efficiency (%) at Temperature of 25° C
INC	80.7	84
P&O	87.8	89.11
MP&O	94.8	94.214

9.5 CONCLUSION

The recommended altered irritate and notice approach is contrasted with the current gradual, annoy and notice strategies. The outcomes show that the changed bother and notice approach, utilized by the PV transfor1mation framework, has a more noteworthy transformation proficiency and tracks the specific most extreme power point significantly quicker with a quicker following rate than the gradual conductance, irritate, and notice strategy. Because of its quicker following pace and high change viability, the altered perturb and observe technique was consequently the one that was generally broadly utilized.

REFERENCES

1. Bidyadhar Subudhi and Raseswar Pradhan. "A Comparative Study on Maximum Power Point Tracking Techniques for Photovoltaic Power Systems", *IEEE Transactions on Sustainable Energy*, vol. 4, no.1, January 2013.
2. Ahmed K. Abdelsalam, Ahmed M. Massoud, Shehab Ahmed, and Prasad Enjeti, "High Performance Adaptive Perturb and Observe MPPT Technique for Photovoltaic Based Microgrids", *Power Electronics, IEEE Transactions*, vol. 26, 2011.
3. S. Arulselvi Durgadevi and S.P. Natarajan, "Study and Implementation of Maximum Power Point Tracking (MPPT) Algorithm for Photovoltaic Systems", *Electrical Energy Systems (ICEES), 1st International Conference*, 2011.

4. Wang Nian Chun, Wu Mei Yue, and Shi Guo Sheng, "Study on Characteristics of Photovoltaic Cells based on MATLAB Simulation", *Power and Energy Engineering Conference Asia-Pacific*, 2011.
5. W.L. Yu, et al. "A DSP-Based Single-Stage Maximum Power Point Tracking PV Inverter", *APEC*, vol. 25, 2010, pp. 948–952.
6. K.N. Hasan, et al. "An Improved Maximum Power Point Tracking Technique for the Photovoltaic Module with Current Mode Control", *AUPEC*, vol. 19, 2009, pp. 1–6.
7. Kalantari, "A Faster Maximum Power Point Tracker Using Peak Current Control", *IEEE Symposium on Industrial Electronics and Applications*, 2009.
8. H. Abouobaida, M. El Khayat, and M. Cherkaoui "Low Cost and High Efficiency Experimental MPPT based on Hill Climbing Approach", *Journal of Electrical Engineering*, vol. 14.
9. T. Esram and P.L. Chapman, "Comparison of Photovoltaic Array Maximum Power Point Tracking Techniques", *IEEE Transactions on Energy Conversion*, vol. 22, no. 2, June, 2007, pp. 439–449.
10. R. Arulmurugan, "Enhanced Maximum Power Tracking for PV Energy System Using New Optimized P&O Algorithm with Cyclic Measurement Tracking Controller".
11. T.L. Kottas, Y.S. Boutalis, and A. D. Karlis, "New Maximum Power Point Tracker for PV Arrays Using Fuzzy Controller in Close Cooperation with Fuzzy Cognitive Network", *IEEE Transactions on Energy Conversion*, vol. 21, no. 3, September, 2006.
12. Joe-Air Jiang, Tsong-Liang Huang, Ying Tung Hsiao and Chia-Hong Chen, "Maximum Power Tracking for Photovoltaic Power Systems", *Tamkang Journal of Science and Engineering*, vol. 8, no. 2, 2005, pp. 147–153 (Pubitemid 40929607)
13. B.M.T. Ho, et al., "Use of System Oscillation to Locate the MPP of PV Panels", *IEEE Power Electronics Letters*, vol. 2, no. 1, 2004, pp. 1–5.
14. H.S. Chung, et al., "A Novel Maximum Power Point Tracking Technique for Solar Panels Using a SEPIC or Cuk Converter", *IEEE Transactions on Power Electronics*, vol. 18, no. 3, 2003, pp. 717–724.
15. D.P. Hohm and M.E. Ropp, "Comparative Study of Maximum Power Point Tracking Algorithms Using an Experimental, Programmable, Maximum Power Point Tracking Test Bed", *Proc. Photovoltaic Specialist Conference*, 2000, pp. 1699–1702.
16. Z. Salameh and D. Taylor, "Step-up Maximum Power Point Tracker for Photovoltaic Arrays", *Solar Energy*, vol. 44, 1990, pp. 57–61.
17. Suman Lata Tripathi and Sanjeevikumar Padmanabhan "Green Energy: Fundamentals, Concepts, and Applications". Scrivener Publishing, Wiley, 2020. DOI: 10.1002/9781119760801
18. S.M. Alghuwainem, "A Close Form Solution for the Maximum Power Operating Point of a solar Cell Array", *Solar Energy Materials and Solar Cells*, vol. 46, pp. 249–257.
19. Suman Lata Tripathi, Parvej Ahmad Alvi, and Umashanker Subramaniam, "Electrical and Electronic Devices, Circuits and Materials: Technological Challenges and Solutions". Scrivener Publishing, Wiley. DOI: 10.1002/9781119755104

Index

Pages in *italics* refer to figures.

Printed in the United States
by Baker & Taylor Publisher Services